# SpringerBriefs in Mathematics

**SpringerBriefs in Mathematics** showcases expositions in all areas of mathematics and applied mathematics. Manuscripts presenting new results or a single new result in a classical field, new field, or an emerging topic, applications, or bridges between new results and already published works, are encouraged. The series is intended for mathematicians and applied mathematicians.

For further volumes:
http://www.springer.com/series/10030

M. Mursaleen

# Applied Summability
# Methods

 Springer

M. Mursaleen
Department of Mathematics
Aligarh Muslim University
Aligarh, Uttar Pradesh, India

ISSN 2191-8198          ISSN 2191-8201 (electronic)
ISBN 978-3-319-04608-2     ISBN 978-3-319-04609-9 (eBook)
DOI 10.1007/978-3-319-04609-9
Springer Cham Heidelberg New York Dordrecht London

Library of Congress Control Number: 2014932267

Mathematics Subject Classification (2010): 40C05, 40G05, 40G10, 40G15, 40H05, 41A36, 42C10, 42A24, 46A45, 47H10

Printed on acid-free paper

Springer is part of Springer Science+Business Media (www.springer.com)

*This book is dedicated to my guide*
*Prof. Z.U. Ahmad from whom I learnt a lot.*

*On the occasion of my marriage anniversary*

# Preface

The theory of summability arises from the process of summation of series and the significance of the concept of summability has been strikingly demonstrated in various contexts, e.g., in analytic continuation, quantum mechanics, probability theory, Fourier analysis, approximation theory, and fixed point theory. The methods of almost summability and statistical summability have become an active area of research in recent years.

This short monograph is the first one to deal exclusively with the study of some summability methods and their interesting applications. We consider here some special regular matrix methods as well as non-matrix methods of summability. This book consists of 12 chapters. In Chap. 1, we recall some basic definitions of sequence spaces, matrix transformations, regular matrices, and some special matrices. Chapter 2 deals with the proof of the prime number theorem by using Lambert's summability and Wiener's Tauberian theorem. In Chap. 3, we give some results on summability tests for singular points of an analytic function. In Chap. 4, we study analytic continuation through Lototski summability. In Chap. 5, we give application of summability methods to independent identically distributed random variables. In Chap. 6, we study a non-matrix method of summability, i.e., almost summability which is further applied in Chaps. 7 and 8 to study the summability of Taylor series, Fourier series, and Walsh-Fourier series. We further use almost summability in Chap. 9 to prove Korovkin type approximation theorems. In Chap. 10, we study another non-matrix method of summability, i.e., statistical summability. In Chap. 11, we study statistical approximation, and in the last chapter, we give some applications of summability methods in fixed point theorems. For the convenience of readers, all chapters of this book are written in a self-contained style and all necessary background and motivations are given per chapter. As such this brief monograph is suitable for researchers, graduate students, and seminars on the above subject.

The author is very much thankful to all three learned referees for their valuable and helpful suggestions.

The author would also like to thank his family for moral support during the preparation of this monograph.

Aligarh, India                                                                    M. Mursaleen
October 15, 2013

# Contents

# Chapter 1
# Toeplitz Matrices

## 1.1 Introduction

The theory of matrix transformations deals with establishing necessary and sufficient conditions on the entries of a matrix to map a sequence space $X$ into a sequence space $Y$. This is a natural generalization of the problem to characterize all summability methods given by infinite matrices that preserve convergence.

In this chapter, we shall present some important classes of matrices such as conservative and regular matrices and enlist some important and very useful special summability matrices. In the subsequent chapters some of these matrices will be used to demonstrate their interesting applications.

## 1.2 Definitions and Notations

### 1.2.1 Classical Sequence Spaces

We denote by $\omega$ the space of all sequences $x = (x_k)_{k=1}^{\infty}$ real or complex, by $\phi$ we denote the set of all finite sequences, that is, sequences which have a finite number of nonzero terms, and write $\ell_\infty$, $c$, $c_0$, and $\ell_p$ for the classical sequence spaces of all bounded, convergent, null, and absolutely $p$-summable sequences of complex numbers, respectively, where $0 < p < \infty$. Also by bs and cs, we denote the spaces of all bounded and convergent series, respectively. $bv_1$ and $bv$ are the spaces of all sequences of bounded variation, that is, consisting of all sequences $(x_k)$ such that $(x_k - x_{k-1})$ and $(x_k - x_{k+1})$ in $\ell_1$, respectively, and $bv_0$ is the intersection of the spaces $bv$ and $c_0$. Let $e = (1, 1, \ldots)$ and $e^{(k)} = (0, 0, \ldots, 0, 1 (k\text{th place}), 0, \ldots)$.

The most popular metric on the space $\omega$ is defined by

$$d_\omega(x, y) = \sum_{k=0}^{\infty} \frac{|x_k - y_k|}{2^k (1 + |x_k - y_k|)}; \quad x = (x_k), \ y = (y_k) \in \omega.$$

M. Mursaleen, *Applied Summability Methods*, SpringerBriefs in Mathematics, DOI 10.1007/978-3-319-04609-9_1, © M. Mursaleen 2014

The space $\ell_\infty$ of bounded sequences is defined by

$$\ell_\infty := \left\{ x = (x_k) \in \omega : \sup_{k \in \mathbb{N}} |x_k| < \infty \right\}.$$

The natural metric on the space $\ell_\infty$ is defined by

$$d_\infty(x, y) = \sup_{k \in \mathbb{N}} |x_k - y_k|; \quad x = (x_k), \ y = (y_k) \in \ell_\infty.$$

The spaces $c$ and $c_0$ of convergent and null sequences are given by

$$c := \left\{ x = (x_k) \in \omega : \lim_{k \to \infty} |x_k - l| = 0 \text{ for some } l \in \mathbb{C} \right\},$$

$$c_0 := \left\{ x = (x_k) \in \omega : \lim_{k \to \infty} x_k = 0 \right\}.$$

The metric $d_\infty$ is also a metric for the spaces $c$ and $c_0$.

The space $\ell_1$ of absolutely convergent series is defined as

$$\ell_1 := \left\{ x = (x_k) \in \omega : \sum_{k=0}^{\infty} |x_k| < \infty \right\}.$$

The space $\ell_p$ of absolutely $p$-summable sequences is defined as

$$\ell_p := \left\{ x = (x_k) \in \omega : \sum_{k=0}^{\infty} |x_k|^p < \infty \right\}, \quad (0 < p < \infty).$$

In the case $1 \le p < \infty$, the metric $d_p$ on the space $\ell_p$ is given by

$$d_p(x, y) = \left( \sum_{k=0}^{\infty} |x_k - y_k|^p \right)^{1/p}; \quad x = (x_k), \ y = (y_k) \in \ell_p.$$

Also in the case $0 < p < 1$, the metric $\tilde{d}_p$ on the space $\ell_p$ is given by

$$\tilde{d}_p(x, y) = \sum_{k=0}^{\infty} |x_k - y_k|^p; \quad x = (x_k), \ y = (y_k) \in \ell_p.$$

The space bs of bounded series is defined by

$$\text{bs} := \left\{ x = (x_k) \in \omega : \sup_{n \in \mathbb{N}} \left| \sum_{k=0}^{n} x_k \right| < \infty \right\}.$$

The natural metric on the space bs is defined by

$$d(x, y) = \sup_{n \in \mathbb{N}} \left| \sum_{k=0}^{n} (x_k - y_k) \right|; \quad x = (x_k), \ y = (y_k) \in \text{bs}. \qquad (1.2.1)$$

The space cs of convergent series and the space $cs_0$ of the series converging to zero are defined as follows:

$$cs := \left\{ x = (x_k) \in \omega : \lim_{n \to \infty} \left| \sum_{k=0}^{n} x_k - l \right| = 0 \text{ for some } l \in \mathbb{C} \right\},$$

$$cs_0 := \left\{ x = (x_k) \in \omega : \lim_{n \to \infty} \left| \sum_{k=0}^{n} x_k \right| = 0 \right\}.$$

The metric $d$ defined by (1.2.1) is the natural metric on the spaces cs and $cs_0$.

The space bv of sequences of bounded variation is defined by

$$bv := \left\{ x = (x_k) \in \omega : \sum_{k=0}^{\infty} |x_k - x_{k+1}| < \infty \right\}.$$

Define the difference sequence $\Delta u = \{(\Delta u)_k\}$ by $(\Delta u)_k = u_k - u_{k+1}$ for all $k \in \mathbb{N}$ with $u_{-1} = 0$. The natural metric on the space $bv_1$ is defined by

$$d(x, y) = |\lim(x-y)| + \sum_{k=0}^{\infty} |[\Delta(x-y)]_k|; \quad x=(x_k), \ y=(y_k) \in bv. \qquad (1.2.2)$$

## 1.2.2  β-Dual

The $\beta$-dual or ordinary Köthe-Toeplitz dual of $X$ is defined by

$$X^{\beta} := \left\{ a = (a_k) \in \omega : \sum_{k=0}^{\infty} a_k x_k \text{ converges for all } x \in X \right\}.$$

Note that $\ell_{\infty}^{\beta} = c_0^{\beta} = c^{\beta} = \ell_1$, $\ell_1^{\beta} = \ell_{\infty}$, $\ell_p^{\beta} = \ell_q$ ($1 < p, q < \infty$, with $p^{-1} + q^{-1} = 1$), $cs^{\beta} = bv$, $bv^{\beta} = cs$, $\omega^{\beta} = \phi$.

## 1.2.3  Schauder Basis

A Schauder basis or countable basis is similar to the usual (Hamel) basis of a vector space; the difference is that Hamel bases use linear combinations that are finite sums, while for Schauder bases they may be infinite sums. This makes Schauder bases more suitable for the analysis of infinite-dimensional topological vector spaces including Banach spaces. A Hamel basis is free from topology while a Schauder basis depends on the metric in question since it involves the notion of "convergence" in its definition and hence topology.

A sequence $(b_k)_{k=0}^\infty$ in a linear metric space $(X, d)$ is called a *Schauder basis* (or briefly *basis*) for $X$ (cf. [59]) if for every $x \in X$ there exists a unique sequence $(\alpha_k)_{k=0}^\infty$ of scalars such that $x = \sum_{k=1}^\infty \alpha_k b_k$, that is, $d(x, x^{[n]}) \to 0$ $(n \to \infty)$, where $x^{[n]} = \sum_{k=0}^n \alpha_k b_k$ is known as the *n*-section of $x$. The series $\sum_{k=0}^\infty \alpha_k b_k$ which has the sum $x$ is called the *expansion* of $x$, and $(\alpha_k)$ is called the *sequence of coefficients* of $x$ with respect to the basis $(b_k)$.

*Example 1.2.1.* The following statements hold:

  (i) The space $\ell_\infty$ has no Schauder basis, since it is not separable.
 (ii) The spaces $\omega$, $c_0$, and $\ell_p$ $(1 \le p < \infty)$ have $(e^{(k)})_{k=1}^\infty$ as their Schauder bases.
(iii) We put $b^{(0)} = e$ and $b^{(k)} = e^{(k-1)}$ for $k = 1, 2, \ldots$ Then the sequence $(b^{(k)})_{k=0}^\infty$ is a Schauder basis for $c$. More precisely, every sequence $x \in c$ has a unique representation $x = le + \sum_{k=0}^\infty (x_k - l)e^{(k)}$ where $l = \lim_{k \to \infty} x_k$.

### 1.2.4   Matrix Transformation

If $A$ is an infinite matrix with complex entries $a_{nk}$ $(n, k \in \mathbb{N})$, then we may write $A = (a_{nk})$ instead of $A = (a_{nk})_{n,k=0}^\infty$. Also, we write $A_n$ for the sequence in the *n*th row of $A$, i.e., $A_n = (a_{nk})_{k=0}^\infty$ for every $n \in \mathbb{N}$. In addition, if $x = (x_k) \in \omega$, then we define the *A-transform* of $x$ as the sequence $Ax = \{A_n(x)\}_{n=0}^\infty$, where

$$A_n(x) = \sum_{k=0}^\infty a_{nk} x_k \ (n \in \mathbb{N})$$

provided the series on the right converges for each $n \in \mathbb{N}$. Further, the sequence $x$ is said to be *A-summable* to the complex number $l$ if $A_n(x) \to l$, as $n \to \infty$, we shall write $x \to l(A)$, where $l$ is called the *A-limit of x*.

Let $X$ and $Y$ be subsets of $\omega$ and $A$ an infinite matrix. Then, we say that $A$ defines a *matrix mapping* from $X$ into $Y$ if $Ax$ exists and is in $Y$ for every $x \in X$. By $(X, Y)$, we denote the class of all infinite matrices that map $X$ into $Y$. Thus $A \in (X, Y)$ if and only if $A_n \in X^\beta$ for all $n \in \mathbb{N}$ and $Ax \in Y$ for all $x \in X$.

### 1.2.5   Continuous Dual

Let $X$ and $Y$ be normed linear spaces. Then $\mathcal{B}(X, Y)$ denotes the set of all bounded linear operators $L : X \to Y$. If $Y$ is complete, then $\mathcal{B}(X, Y)$ is a Banach space with the operator norm defined by $\|L\| = \sup_{x \in S_X} \|L(x)\|$ for all $L \in \mathcal{B}(X, Y)$. By $X' = \mathcal{B}(X, \mathbb{C})$, we denote the *continuous dual* of $X$, that is, the set of all continuous linear functionals on $X$. If $X$ is a Banach space, then we write $X^*$ for $X'$ with its norm given by $\|f\| = \sup_{x \in S_X} |f(x)|$ for all $f \in X'$, where $S_X$ is the unit sphere in $X$.

## 1.3 Conservative and Regular Matrices

It was the celebrated German mathematician Otto Toeplitz (1881–1940) who characterized those matrices $A = (a_{nk})$ which transform convergent sequences into convergent sequences leaving the limit invariant (see [13, 17, 26, 41, 59]). A summability method is an alternative formulation of convergence of a series which is divergent in the conventional sense.

**Definition 1.3.1.** A matrix $A$ is called a *conservative matrix* if $Ax \in c$ for all $x \in c$. If in addition $\lim Ax = \lim x$ for all $x \in c$, then $A$ is called a *regular matrix or regular method* or *Toeplitz matrix*. The class of conservative matrices will be denoted by $(c, c)$ and of regular matrices by $(c, c; P)$ or $(c, c)_{\text{reg}}$.

**Definition 1.3.2.** A matrix $A$ is called a *Schur matrix* or *coercive matrix* if $Ax \in c$ for all $x \in \ell_\infty$. The class of Schur matrices will be denoted by $(\ell_\infty, c)$.

**Theorem 1.3.3 (Silverman-Toeplitz theorem).** $A = (a_{nk}) \in (c, c; P)$ *if and only if*

$$\|A\| = \sup_{n \in \mathbb{N}} \sum_{k=0}^{\infty} |a_{nk}| < \infty, \tag{1.3.1}$$

$$\lim_{n \to \infty} a_{nk} = 0 \text{ for each } k \in \mathbb{N}, \tag{1.3.2}$$

$$\lim_{n \to \infty} \sum_{k=0}^{\infty} a_{nk} = 1. \tag{1.3.3}$$

The following is the more general class.

**Theorem 1.3.4 (Kojima-Schur).** $A = (a_{nk}) \in (c, c)$ *if and only if (1.3.1) holds and there exist $\alpha_k, \alpha \in \mathbb{C}$ such that*

$$\lim_{n \to \infty} a_{nk} = \alpha_k \text{ for each } k \in \mathbb{N}, \tag{1.3.4}$$

$$\lim_{n \to \infty} \sum_{k=0}^{\infty} a_{nk} = \alpha. \tag{1.3.5}$$

If $A \in (c, c)$ and $x \in c$, then

$$\lim_{n \to \infty} A_n(x) = \left( \alpha - \sum_{k=0}^{\infty} \alpha_k \right) \lim_{k \to \infty} x_k + \sum_{k=0}^{\infty} \alpha_k x_k. \tag{1.3.6}$$

*Proof.* Suppose that the conditions (1.3.1), (1.3.4), and (1.3.5) hold and $x = (x_k) \in c$ with $x_k \to l$ as $k \to \infty$. Then, since $(a_{nk})_{k \in \mathbb{N}} \in c^\beta = \ell_1$ for each $n \in \mathbb{N}$, the $A$-transform of $x$ exists. In this case, the equality

$$\sum_{k=0}^{\infty} a_{nk} x_k = \sum_{k=0}^{\infty} a_{nk} (x_k - l) + l \sum_{k=0}^{\infty} a_{nk} \tag{1.3.7}$$

holds for each $n \in \mathbb{N}$. In (1.3.7), since the first term on the right-hand side tends to $\sum_{k=0}^{\infty} \alpha_k (x_k - l)$ by (1.3.4) and the second term on the right-hand side tends to $l\alpha$ by (1.3.5) as $n \to \infty$, we have

$$\lim_{n \to \infty} \sum_{k=0}^{\infty} a_{nk} x_k = \sum_{k=0}^{\infty} \alpha_k (x_k - l) + l\alpha. \tag{1.3.8}$$

Hence, $Ax \in c$, that is, $A \in (c, c)$.

Conversely, suppose that $A \in (c, c)$. Then $Ax$ exists for every $x \in c$. The necessity of the conditions (1.3.4) and (1.3.5) is immediate by taking $x = e^{(k)}$ and $x = e$, respectively. Now, using the Banach-Steinhaus theorem and the closed graph theorem, we have $A \in \mathcal{B}(c, c)$. Thus,

$$\sup_{n \in \mathbb{N}} \left| \sum_{k=0}^{\infty} a_{nk} x_k \right| \le \|A\| \|x\|_\infty \tag{1.3.9}$$

for all $x \in c$. Now choose any $n \in \mathbb{N}$ and any $r \in \mathbb{N}$ and define $x \in c_0$ by

$$x_k = \begin{cases} \operatorname{sgn} a_{nk}, & 1 \le k \le r, \\ 0, & k > r. \end{cases}$$

Substituting this in (1.3.9) we get

$$\sum_{k=1}^{r} |a_{nk}| \le \|A\|. \tag{1.3.10}$$

Letting $r \to \infty$ and noting that (1.3.10) holds for every $n \in \mathbb{N}$ we observe that (1.3.1) holds. Finally, (1.3.8) is same as (1.3.6).

This completes the proof. □

*Remark 1.3.5.* Taking $\alpha_k = 0$ for all $k \in \mathbb{N}$ and $\alpha = 1$ in Theorem 1.3.4, we get Theorem 1.3.3.

First we state the following lemma which is needed in proving Schur's theorem.

**Lemma 1.3.6.** *Let $B = (b_{nk})_{n,k}$ be an infinite matrix such that $\sum_k |b_{nk}| < \infty$ for each $n$ and $\sum_k |b_{nk}| \to 0$ $(n \to \infty)$. Then $\sum_k |b_{nk}|$ converges uniformly in $n$.*

*Proof.* $\sum_k |b_{nk}| \to 0$ $(n \to \infty)$ implies that $\sum_k |b_{nk}| < \infty$ for $n \ge N(\varepsilon)$. Since $\sum_k |b_{nk}| < \infty$ for $0 \le n \le N(\varepsilon)$, there exists $m = M(\varepsilon, n)$ such that $\sum_{k \ge M} |b_{nk}| < \infty$ for all $n$, which means that $\sum_k |b_{nk}|$ converges uniformly in $n$.

This completes the proof of the lemma. □

**Theorem 1.3.7 (Schur).** *$A = (a_{nk}) \in (\ell_\infty, c)$ if and only if (1.3.4) holds and*

$$\sum_{k=0}^{\infty} |a_{nk}| \text{ converges uniformly in } n \in \mathbb{N}. \tag{1.3.11}$$

*Proof.* Suppose that the conditions (1.3.4) and (1.3.11) hold and $x \in \ell_\infty$. Then, $\sum_k a_{nk} x_k$ is absolutely and uniformly convergent in $n \in \mathbb{N}$. Hence, $\sum_k a_{nk} x_k \to \sum_k \alpha_k x_k$ $(n \to \infty)$ which gives that $A \in (\ell_\infty, c)$.

Conversely, suppose that $A \in (\ell_\infty, c)$ and $x \in \ell_\infty$. Then necessity of (1.3.4) follows easily by taking $x = e^{(k)}$ for each $k$. Define $b_{nk} = a_{nk} - \alpha_k$ for all $k, n \in \mathbb{N}$. Since $\sum_k |\alpha_k| < \infty$, $(\sum_k b_{nk} x_k)_n$ converges whenever $x = (x_k) \in \ell_\infty$. Now if we can show that this implies

$$\lim_n \sum_k |b_{nk}| = 0, \tag{1.3.12}$$

then by using Lemma 1.3.6, we shall get the desired result. Suppose to the contrary that $\lim_n \sum_k |b_{nk}| \neq 0$. Then, it follows that $\lim_n \sum_k |b_{nk}| = l > 0$ through some subsequence of the positive integers. Also we have $b_{mk} \to 0$ as $m \to \infty$ for each $k \in \mathbb{N}$. Hence we may determine $m(1)$ such that

$$| \sum_k |b_{m(1),k}| - l | < l/10 \text{ and } b_{m(1),1} < l/10.$$

Since $\sum_k |b_{m(1),k}| < \infty$ we may choose $k(2) > 1$ such that

$$\sum_{k=k(2)+1}^{\infty} |b_{m(1),k}| < l/10.$$

It follows that

$$| \sum_{k=2}^{k(2)} |b_{m(1),k}| - l | < l/10.$$

For our convenience we use the notation $\sum_{k=p}^q |b_{mk}| = B(m, p, q)$.

Now we choose $m(2) > m(1)$ such that $| B(m(2), 1, \infty) - l | < l/10$ and $B(m(2), 1, k(2)) < l/10$. Then choose $k(3) > k(2)$ such that $| B(m(2), k(3) + 1, \infty) - l | < l/10$. It follows that $| B(m(2), k(2)+1, k(3)) - l | < 3l/10$. Continuing in this way and find $m(1) < m(2) < \dots$, $1 = k(1) < k(2) < \dots$ such that

$$\begin{cases} B(m(r), 1, k(r)) < l/10 \\ B(m(r), k(r+1) + 1, \infty) < l/10 \\ B(m(r), k(r)+1, k(r+1)) - l | < 3l/10. \end{cases} \tag{1.3.13}$$

Let us define $x = (x_k) \in \ell_\infty$ such that $\| x \| = 1$ by

$$x_k = \begin{cases} 0, & \text{if } k = 1, \\ (-1)^r \operatorname{sgn}(b_{m(r),k}), & \text{if } k(r) < k \le k(r+1), \end{cases} \tag{1.3.14}$$

for $r = 1, 2, \ldots$. Then write $\sum_k b_{m(r),k} x_k$ as $\sum_1 + \sum_2 + \sum_3$, where $\sum_1$ is over $1 \leq k \leq k(r)$, $\sum_2$ is over $k(r) < k \leq k(r+1)$, and $\sum_2$ is over $k > k(r+1)$. It follows immediately from (1.3.13) with the sequence $x$ given by (1.3.14) that

$$\left| \sum_k b_{m(r),k} - (-1)^r l \right| < l/2.$$

Consequently, it is clear that the sequence $Bx = (\sum_k b_{nk} x_k)$ is not a Cauchy sequence and so is not convergent. Thus we have proved that $Bx$ is not convergent for all $x \in \ell_\infty$ which contradicts the fact that $A \in (\ell_\infty, c)$. Hence, (1.3.12) must hold. Now, it follows by Lemma 1.3.6 that $\sum_k |b_{nk}|$ converges uniformly in $n$. Therefore, $\sum_k |a_{nk}| = \sum_k |b_{nk} + \alpha_k|$ converges uniformly in $n$.

This completes the proof.                                                                 □

We get the following corollary:

**Corollary 1.3.8.** $A \in (\ell_\infty, c_0)$ if and only if

$$\lim_n \sum_k |b_{nk}| = 0. \tag{1.3.15}$$

We observe the following application of Corollary 1.3.8.

**Theorem 1.3.9.** *Weak and strong convergence coincide in $\ell_1$.*

*Proof.* We assume that the sequence $(x^{(n)})_{n=0}^\infty$ is weakly convergent to $x$ in $\ell_1$, that is, $|f(x^{(n)}) - f(x)| \to 0$ $(n \to \infty)$ for every $f \in \ell_1^*$. Since $\ell_1^*$ and $\ell_\infty$ are norm isomorphic, to every $f \in \ell_1^*$ there corresponds a sequence $a \in \ell_\infty$ such that $f(y) = \sum_{k=0}^\infty a_k y_k$. We define the matrix $B = (b_{nk})_{n,k=0}^\infty$ by $b_{nk} = x_k^{(n)} - x_k$ $(n, k = 0, 1, \ldots)$. Then we have $f(x^{(n)}) - f(x) = \sum_{k=0}^\infty a_k (x_k^{(n)} - x_k) = \sum_{k=0}^\infty b_{nk} a_k \to 0$ $(n \to \infty)$ for all $a \in \ell_\infty$, that is, $B \in (\ell_\infty, c_0)$, and it follows from Corollary 1.3.8 that $\|x^{(n)} - x\|_{\ell_1} = \sum_{k=0}^\infty |x_k^{(n)} - x_k| = \sum_{k=0}^\infty |b_{nk}| \to 0$ $(n \to \infty)$.

This completes the proof of the theorem.                                                 □

**Definition 1.3.10.** The *characteristic* $\chi(A)$ of a matrix $A = (a_{nk}) \in (c, c)$ is defined by

$$\chi(A) = \lim_{n \to \infty} \sum_{k=0}^\infty a_{nk} - \sum_{k=0}^\infty \left( \lim_{n \to \infty} a_{nk} \right)$$

which is a multiplicative linear functional. The numbers $\lim_{n \to \infty} a_{nk}$ and $\lim_{n \to \infty} \sum_{k=0}^\infty a_{nk}$ are called the *characteristic numbers* of $A$. A matrix $A$ is called *coregular* if $\chi(A) \neq 0$ and is called *conull* if $\chi(A) = 0$.

*Remark 1.3.11.* The Silverman-Toeplitz theorem yields for a regular matrix $A$ that $\chi(A) = 1$ which leads us to the fact that regular matrices form a subset of coregular matrices. One can easily see for a Schur matrix $A$ that $\chi(A) = 0$ which tells us that coercive matrices form a subset of conull matrices. Hence we have the following result which is known as *Steinhaus's theorem*.

**Theorem 1.3.12 (Steinhaus).** *For every regular matrix $A$, there is a bounded sequence which is not $A$-summable.*

*Proof.* We assume that a matrix $A \in (c, c; P) \cap (\ell_\infty, c)$. Then it follows from Theorem 1.3.3 and Schur's theorem that $1 = \lim_{n\to\infty} \sum_{k=0}^\infty a_{nk} = \sum_{k=0}^\infty (\lim_{n\to\infty} a_{nk}) = 0$, a contradiction.

This completes the proof.                                                                   □

## 1.4  Some Special Summability Matrices

First we give here some special and important matrices of triangles. The most important summability methods are given by Hausdorff matrices and their special cases.

(i) **Hausdorff Matrix.** Let $\mu = (\mu_n)_{n=0}^\infty$ be a given complex sequence, $M = (m_{nk})_{n,k=0}^\infty$ be the diagonal matrix with $m_{nn} = \mu_n$ $(n = 0, 1, \ldots)$, and $D = (d_{nk})_{n,k=0}^\infty$ be the matrix with $d_{nk} = (-1)^k \binom{n}{k}$. Then the matrix $H = H(\mu) = DMD$ is called the *Hausdorff matrix associated with the sequence $\mu$*; i.e.,

$$h_{nk} = \begin{cases} \sum_{j=k}^n (-1)^{j+k} \binom{n}{j}\binom{j}{k} , & 0 \le k \le n, \\ 0 & , k > n, \end{cases}$$

for all $k, n \in \mathbb{N}_0$.

(ii) **Cesàro Matrix.** The *Cesàro matrix of order* 1 is defined by the following matrix $C_1 = (c_{nk})$

$$c_{nk} = \begin{cases} \frac{1}{n+1} , & 0 \le k \le n, \\ 0 & , k > n. \end{cases}$$

The inverse matrix $C_1^{-1} = (d_{nk})$ of the matrix $C_1 = (c_{nk})$ is given by

$$d_{nk} = \begin{cases} (-1)^{n-k}(k+1) , & n-1 \le k \le n, \\ 0 & , 0 \le n \le n-2 \text{ or } k > n, \end{cases}$$

for all $k, n \in \mathbb{N}_0$.

Let $r > -1$ and define $A_n^r$ by

$$A_n^r = \begin{cases} \frac{(r+1)(r+2)\cdots(r+n)}{n!} , & n = 1, 2, \ldots, \\ 1 & , n = 0. \end{cases}$$

Then the *Cesàro matrix of order r* is defined by the following matrix $C_r = (c_{nk}^r)$

$$c_{nk}^r = \begin{cases} \frac{A_{n-k}^{r-1}}{A_n^r} & , 0 \le k \le n, \\ 0 & , k > n, \end{cases}$$

for all $k, n \in \mathbb{N}_0$.

(iii) **Euler Matrix.** The *Euler matrix* $E_1$ of order 1 is given by the matrix $E_1 = (a_{nk})$, where

$$a_{nk} = \begin{cases} \binom{n}{k} 2^{-n} & , 0 \le k \le n, \\ 0 & , k > n, \end{cases}$$

for all $k, n \in \mathbb{N}_0$ whose generalization $E_q$ of order $q > 0$ was defined by the matrix $E_q = (b_{nk}^q)$, where

$$b_{nk}^q = \begin{cases} \binom{n}{k}(q+1)^{-n}q^{n-k} & , 0 \le k \le n, \\ 0 & , k > n, \end{cases}$$

for all $k, n \in \mathbb{N}_0$.

Let $0 < r < 1$ and $\binom{n}{k} = n!/[k!(n-k)!]$ for all $k, n \in \mathbb{N}_0$. Then the *Euler matrix* $E^r$ *of order r* is defined by the matrix $E^r = (e_{nk}^r)$, where

$$e_{nk}^r = \begin{cases} \binom{n}{k}(1-r)^{n-k}r^k & , 0 \le k \le n, \\ 0 & , k > n, \end{cases}$$

for all $k, n \in \mathbb{N}_0$. It is clear that $E^r$ corresponds to $E_q$ for $r = (q+1)^{-1}$. Much of the work on the Euler means of order $r$ was done by Knopp [53]. So, some authors refer to $E^r$ as the *Euler-Knopp matrix*. The original Euler means $E_1 = E^{1/2}$ was given by L. Euler in 1755. $E^r$ is invertible such that $(E^r)^{-1} = E^{1/r}$ with $r \ne 0$.

(iv) **Riesz Matrix.** Let $t = (t_k)$ be a sequence of nonnegative real numbers with $t_0 > 0$ and write $T_n = \sum_{k=0}^n t_k$ for all $n \in \mathbb{N}_0$. Then the *Riesz matrix with respect to the sequence* $t = (t_k)$ is defined by the matrix $R^t = (r_{nk}^t)$ which is given by

$$r_{nk}^t = \begin{cases} \frac{t_k}{T_n} & , 0 \le k \le n, \\ 0 & , k > n, \end{cases}$$

for all $k, n \in \mathbb{N}_0$. For $t = e$ the Riesz matrix $R^t$ is reduced to the matrix $C_1$. The inverse matrix $S^t = (s_{nk}^t)$ of the matrix $R^t = (r_{nk}^t)$ is given by

$$s_{nk}^t = \begin{cases} \frac{(-1)^{n-k}T_k}{t_n} & , n-1 \le k \le n, \\ 0 & , 0 \le k \le n-2 \text{ or } k > n. \end{cases}$$

(v) **Nörlund Matrix.** Let $q = (q_k)$ be a sequence of nonnegative real numbers with $q_0 > 0$ and write $Q_n = \sum_{k=0}^{n} q_k$ for all $n \in \mathbb{N}_0$. Then the *Nörlund matrix with respect to the sequence* $q = (q_k)$ is defined by the matrix $N^q = (a_{nk}^q)$ which is given by

$$a_{nk}^q = \begin{cases} \frac{q_{n-k}}{Q_n} , 0 \leq k \leq n, \\ 0 , k > n, \end{cases}$$

for all $k, n \in \mathbb{N}_0$. For $q = e$ the Nörlund matrix $N^q$ is reduced to the matrix $C_1$. Now, write $t(z) = \sum_n t_n z^n$, $k(z) = 1/t(z) = \sum_n k_n z^n$. The inverse $M^q$ of $N^q$ is then given by $(M^q)_{nj} = k_{n-j} Q_j$ for $j \leq n$ (cf. Peyerimhoff [80, p. 17]). In the case $t_n = A_n^{r-1}$ for all $n \in \mathbb{N}_0$, the method $N^q$ is reduced to the Cesàro method $C_r$ of order $r > -1$.

(vi) **Borel Matrix.** The *Borel matrix* $B = (b_{nk})_{n,k=1}^{\infty}$ is defined by

$$b_{nk} = e^{-n} n^k / k!$$

for all $k, n \in \mathbb{N}_0$.

*Remark 1.4.1.* The following statements hold:

1. The Cesàro matrix of order $r$ is a Toeplitz matrix if $r \geq 0$.
2. The Euler matrix $E^r$ of order $r$ is a Toeplitz matrix if and only if $0 < r \leq 1$.
3. The Riesz matrix $R^t$ is a Toeplitz matrix if and only if $T_n \to 0$ as $n \to \infty$.
4. The Nörlund matrix $N^q$ is a Toeplitz matrix if and only if $q_n/Q_n \to 0$ as $n \to \infty$.
5. The Borel matrix is a Toeplitz matrix.

# Chapter 2
# Lambert Summability and the Prime Number Theorem

## 2.1 Introduction

The *prime number theorem* (PNT) was stated as conjecture by German mathematician Carl Friedrich Gauss (1777–1855) in the year 1792 and proved independently for the first time by Jacques Hadamard and Charles Jean de la Vallée-Poussin in the same year 1896. The first elementary proof of this theorem (without using integral calculus) was given by Atle Selberg of Syracuse University in October 1948. Another elementary proof of this theorem was given by Erdös in 1949.

The PNT describes the asymptotic distribution of the prime numbers. The PNT gives a general description of how the primes are distributed among the positive integers.

Informally speaking, the PNT states that if a random integer is selected in the range of zero to some large integer $N$, the probability that the selected integer is prime is about $1/\ln(N)$, where $\ln(N)$ is the natural logarithm of $N$. For example, among the positive integers up to and including $N = 10^3$, about one in seven numbers is prime, whereas up to and including $N = 10^{10}$, about one in 23 numbers is prime (where $\ln(103) = 6.90775528$ and $\ln(1010) = 23.0258509$). In other words, the average gap between consecutive prime numbers among the first $N$ integers is roughly $\ln(N)$.

Here we give the proof of this theorem by the application of Lambert summability and Wiener's Tauberian theorem. The Lambert summability is due to German mathematician Johann Heinrich Lambert (1728–1777) (see Hardy [41, p. 372]; Peyerimhoff [80, p. 82]; Saifi [86]).

## 2.2 Definitions and Notations

(i) **Möbius Function.** The classical *Möbius function* $\mu(n)$ is an important multiplicative function in number theory and combinatorics. This formula is due to German mathematician August Ferdinand Möbius (1790–1868) who

M. Mursaleen, *Applied Summability Methods*, SpringerBriefs in Mathematics, DOI 10.1007/978-3-319-04609-9_2, © M. Mursaleen 2014

introduced it in 1832. $\mu(n)$ is defined for all positive integers $n$ and has its values in $\{-1, 0, 1\}$ depending on the factorization of $n$ into prime factors. It is defined as follows (see Peyerimhoff [80, p. 85]):

$$\mu(n) = \begin{cases} 1 & , n \text{ is a square-free positive integer with an even number of prime factors,} \\ -1 & , n \text{ is a square-free positive integer with an odd number of prime factors,} \\ 0 & , n \text{ is not square-free,} \end{cases}$$

that is,

$$\mu(n) = \begin{cases} 1 & , n = 1, \\ (-1)^k & , n = p_1 p_2 \cdots p_k, \ p_i \text{ prime, } p_i \neq p_j, \\ 0 & , \text{ otherwise.} \end{cases} \qquad (2.2.1)$$

Thus

(a) $\mu(2) = -1$, since $2 = 2$;
(b) $\mu(10) = 1$, since $10 = 2 \times 5$;
(c) $\mu(4) = 0$, since $4 = 2 \times 2$.

We conclude that $\mu(p) = -1$, if $p$ is a prime number.

(ii) **The Function $\pi(x)$.** The *prime-counting function* $\pi(x)$ is defined as the number of primes not greater than $x$, for any real number $x$, that is, $\pi(x) = \sum_{p < x} 1$ (Peyerimhoff [80, p. 87]). For example, $\pi(10) = 4$ because there are four prime numbers (2, 3, 5, and 7) less than or equal to 10. Similarly, $\pi(1) = 0, \pi(2) = 0, \pi(3) = 1, \pi(4) = 2, \pi(1000) = 168, \pi(10^6) = 78498$, and $\pi(10^9) = 50847478$ (Hardy [43, p. 9]).

(iii) **The von Mangoldt Function $\Lambda_n$.** The function $\Lambda_n$ is defined as follows (Peryerimhoff [80, p. 84]):

$$\Lambda_n = \begin{cases} \log p & , n = p^\alpha \text{ for some prime } p \text{ and } \alpha \geq 1, \\ 0 & , \text{ otherwise.} \end{cases}$$

(iv) **Lambert Summability.** A series $\sum_{n=1}^{\infty} a_n$ is said to be *Lambert summable* (or summable $\mathcal{L}$) to $s$, if

$$\lim_{x \to 1^-} (1 - x) \sum_{k=1}^{\infty} \frac{k a_k x^k}{1 - x^k} = s. \qquad (2.2.2)$$

In this case, we write $\sum a_n = s(\mathcal{L})$. Note that if a series is convergent to $s$, then it is Lambert summable to $s$.

This series is convergent for $|x| < 1$, which is true if and only if $a_n = O((1 + \varepsilon)^n)$, for every $\varepsilon > 0$ (see [6, 52, 99]).

If we write $x = e^{-\frac{1}{y}}$ $(y > 0)$, $s(t) = \sum_{k \le t} a_k$ $(a_0 = 0)$, $g(t) = \frac{te^{-t}}{1-e^{-t}}$, then $\sum a_k$ is summable $\mathcal{L}$ to $s$ if and only if (note that $1 - x \approx \frac{1}{y}$)

$$\lim_{y \to \infty} \frac{1}{y} \int_0^\infty \frac{te^{-\frac{t}{y}}}{1 - e^{-\frac{t}{y}}} ds(t) = \lim_{y \to \infty} - \int_0^\infty s(t) dg\left(\frac{t}{y}\right)$$

$$= \lim_{y \to \infty} -\frac{1}{y} \int_0^\infty g'\left(\frac{t}{y}\right) s(t) dt = s.$$

The method $\mathcal{L}$ is regular.

## 2.3  Lemmas

We need the following lemmas for the proof of the PNT which is stated and proved in the next section. In some cases, Tauberian condition(s) will be used to prove the required claim. The general character of a Tauberian theorem is as follows. The ordinary questions on summability consider two related sequences (or other functions) and ask whether it will be true that one sequence possesses a limit whenever the other possesses a limit, the limits being the same; a Tauberian theorem appears, on the other hand, only if this is untrue, and then asserts that the one sequence possesses a limit provided the other sequence both possesses a limit and satisfies some additional condition restricting its rate of increase. The interest of a Tauberian theorem lies particularly in the character of this additional condition, which takes different forms in different cases.

**Lemma 2.3.1 (Hardy [41, p. 296]; Peyerimhoff [80, p. 80]).** *If $g(t), h(t) \in L(0, \infty)$, and if*

$$\int_0^\infty g(t) t^{ix} dt \neq 0 \; (-\infty < x < \infty), \tag{2.3.1}$$

*then $s(t) = O(1)$ ($s(t)$ real and measurable) and*

$$\lim_{x \to \infty} \frac{1}{x} \int_0^\infty g\left(\frac{t}{x}\right) s(t) dt = 0 \; \text{implies} \; \lim_{x \to \infty} \frac{1}{x} \int_0^\infty h\left(\frac{t}{x}\right) s(t) dt = 0.$$

**Lemma 2.3.2 (Peyerimhoff [80, p. 84]).** *If $n = p_1^{\alpha_1} \cdots p_k^{\alpha_k}$ $(\alpha_i = 1, 2, \ldots, p_i$ prime), then $\sum_{d/n} \Lambda_d = \log n$.*

*Proof.* Since $d$ runs through divisors of $n$ and we have to consider only $d = p_1, p_1^2, \ldots, p_1^{\alpha_1}, \ldots, p_k^{\alpha_k}$, therefore $\sum_{d/n} \Lambda_d = \alpha_1 \log p_1 + \alpha_2 \log p_2 + \cdots + \alpha_k \log p_k = \log n$.

This completes the proof of Lemma 2.3.2.  □

**Lemma 2.3.3 (Peyerimhoff [80, p. 84]).**

$$\sum_{n=1}^{\infty} \frac{(-1)^{n-1}}{n^s} = (1 - 2^{1-s})\zeta(s)(s > 1), \qquad (2.3.2)$$

where $\zeta$ is a Riemann's Zeta function.

*Proof.* We have

$$\sum_{n=1}^{\infty} \frac{(-1)^{n-1}}{n^s} = 1 - \frac{1}{2^s} + \frac{1}{3^s} - \frac{1}{4^s} + \cdots$$

$$= \left(1 + \frac{1}{2^s} + \frac{1}{3^s}\right) - 2\left(\frac{1}{2^s} + \frac{1}{4^s} + \frac{1}{6^s} + \cdots\right)$$

$$= \zeta(s) - \frac{2}{2^s}\left(1 + \frac{1}{2^s} + \frac{1}{3^s} + \cdots\right)$$

$$= \zeta(s) - 2^{1-s}\zeta(s)$$

$$= (1 - 2^{1-s})\zeta(s).$$

This completes the proof of Lemma 2.3.3. □

**Lemma 2.3.4 (Hardy [41, p. 246]).** *If $s > 1$, then*

$$\zeta(s) = \prod_p \frac{p^s}{p^s - 1}. \qquad (2.3.3)$$

**Lemma 2.3.5 (Hardy [43, p. 253]).**

$$-\zeta'(s) = \zeta(s) \sum_{n=1}^{\infty} \frac{\Lambda_n}{n^s} \qquad (2.3.4)$$

*Proof.* From 2.3.3, we have

$$\log \zeta(s) = \sum_p \log \frac{p^s}{p^s - 1}.$$

Differentiating with respect to $s$ and observing that

$$\frac{d}{ds}\left(\log \frac{p^s}{p^s - 1}\right) = -\frac{\log p}{p^s - 1},$$

we obtain

$$-\frac{\zeta'(s)}{\zeta(s)} = \sum_p \frac{\log p}{p^s - 1}. \tag{2.3.5}$$

The differentiation is legitimate because the derived series is uniformly convergent for $s \geq 1 + \delta > 1, \delta > 0$.

We can write (2.3.5) in the form

$$-\frac{\zeta'(s)}{\zeta(s)} = \sum_p \log p \sum_{m=1}^{\infty} p^{-ms}$$

and the double series $\sum\sum p^{-ms} \log p$ is absolutely convergent when $s > 1$. Hence it may be written as

$$\sum_{p,m} p^{-ms} \log p = \sum_{n=0}^{\infty} \Lambda_n n^{-s}.$$

This completes the proof of Lemma 2.3.5. □

**Lemma 2.3.6 (Peyerimhoff [80, p. 84]).** $s_n \to s(\mathcal{L})$, as $n \to \infty$ and $a_n = O\left(\frac{1}{n}\right)$ imply $s_n \to s$, as $n \to \infty$.

*Proof.* We wish to show that $a_n = O(1/n)$ is a Tauberian condition. In order to apply Wiener's theory we must show that (2.3.1) holds. But for $\varepsilon > 0$

$$-\int_0^{\infty} t^{ix+\varepsilon} g'(t) dt = (ix + \varepsilon) \int_0^{\infty} t^{ix+\varepsilon-1} g(t) dt$$

$$= (ix + \varepsilon) \sum_{k=0}^{\infty} \int_0^{\infty} t^{ix+\varepsilon} e^{-(k+1)t} dt$$

$$= (ix + \varepsilon)\Gamma(1 + \varepsilon + ix) \sum_{k=0}^{\infty} \frac{1}{(k+1)^{1+\varepsilon+ix}}$$

i.e.,

$$-\int_0^{\infty} t^{ix} g'(t) dt = \Gamma(1 + ix) \lim_{\varepsilon \to 0} (ix + \varepsilon)\zeta(1 + \varepsilon + ix).$$

This has a simple pole at 1 and is $\neq 0$ on the line $\text{Re} z = 1$. A stronger theorem is true, namely, $\mathcal{L} \subseteq Abel$, i.e., every Lambert summable series is also Abel summable (see [42]), which implies this theorem. For the sake of completeness we give a proof that $\zeta(1 + ix) \neq 0$ for real $x$. The formula (2.3.4) implies $\zeta(1 + ix) \neq 0$.

This completes the proof of Lemma 2.3.6. □

**Lemma 2.3.7 (Peyerimhoff [80, p. 86]).**

$$\sum_{n=1}^{\infty} \frac{\mu(n)}{n} = 0$$

*Proof.* This follows from $O$-Tauberian theorem for Lambert summability, if $\sum_{n=1}^{\infty} \frac{\mu(n)}{n} = O(L)$. But

$$(1-x)\sum_{n=1}^{\infty} \frac{\mu(n)x^n}{1-x^n} = (1-x)\sum_{n=1}^{\infty} \mu(n) \sum_{k=0}^{\infty} x^{n(k+1)}$$

$$= (1-x)\sum_{m=1}^{\infty} \sum_{n/m} \mu(n) = x(1-x).$$

A consequence is (by partial summation)

$$\sum_{k \le n} \mu(k) = O(n) \tag{2.3.6}$$

which follows with the notation

$$m(t) = \sum_{1 \le k \le t} \frac{\mu(k)}{k} \quad \text{from} \quad \sum_{k \le n} \mu(k) = \int_{1-0}^{n} t\, dm(t) = nm(n) - \int_{1}^{n} m(t)dt.$$

This completes the proof of Lemma 2.3.7.                                             $\square$

**Lemma 2.3.8 (Hardy [43, p. 346]).** *Suppose that $c_1, c_2, \ldots,$ is a sequence of numbers such that*

$$C(t) = \sum_{n \le t} c_n$$

*and that $f(t)$ is any function of $t$. Then*

$$\sum_{n \le x} c_n f(n) = \sum_{n \le x-1} C(n)\{f(n) - f(n+1)\} + C(x)f([x]). \tag{2.3.7}$$

*If, in addition, $c_j = 0$ for $j < n_1$ and $f(t)$ has a continuous derivative for $t \ge n_1$, then*

$$\sum_{n \le x} c_n f(n) = C(x)f(x) - \int_{n_1}^{x} C(t)f'(t)dt. \tag{2.3.8}$$

*Proof.* If we write $N = [x]$, the sum on the left of (2.3.7) is

$$C(1)f(1) + \{C(2) - C(1)\}f(2) + \cdots + \{C(N) - C(N-1)\}f(N)$$
$$= C(1)\{f(1) - f(2)\} + \cdots + C(N-1)\{f(N-1) - f(N)\} + C(N)f(N).$$

Since $C(N) = C(x)$, this proves (2.3.7). To deduce (2.3.8), we observe that $C(t) = C(n)$ when $n \leq t < n+1$ and so

$$C(n)[f(n) - f(n+1)] = -\int_n^{n+1} C(t)\, f'(t)dt.$$

Also $C(t) = 0$ when $t < n_1$.

This completes the proof of Lemma 2.3.8. $\qquad\qquad\qquad\qquad\qquad\qquad\square$

**Lemma 2.3.9 (Hardy [43, p. 347]).**

$$\sum_{n \leq x} \frac{1}{n} = \log x + C + O\left(\frac{1}{x}\right),$$

*where $C$ is Euler's constant.*

*Proof.* Put $c_n = 1$ and $f(t) = 1/t$. We have $C(x) = [x]$ and (2.3.8) becomes

$$\sum_{n \leq x} \frac{1}{n} = \frac{[x]}{x} + \int_1^x \frac{[t]}{t^2}dt$$

$$= \log x + C + E,$$

where

$$C = 1 - \int_1^\infty \frac{t - [t]}{t^2}dt$$

is independent of $x$ and

$$E = \int_x^\infty \frac{t - [t]}{t^2}dt - \frac{x - [x]}{x}$$

$$= \int_x^\infty \frac{O(1)}{t^2}dt + O\left(\frac{1}{x}\right)$$

$$= O\left(\frac{1}{x}\right).$$

This completes the proof of Lemma 2.3.9. $\qquad\qquad\qquad\qquad\qquad\qquad\square$

**Lemma 2.3.10 (Peyerimhoff [80, p. 86]).** *If*

$$\chi(x) = \sum_{k \le x} \left[ \psi\left(\frac{x}{k}\right) - \frac{x}{k} + \log\frac{x}{k} + C \right] \text{ and } \psi(x) = \sum_{n \le x} \Lambda_n,$$

*then $\chi(x) = O(\log(x+1))$.*

*Proof.* Möbius formula (2.2.1) yields that

$$\psi(x) - x + \log x + C = \sum_{d \le x} \chi\left(\frac{x}{d}\right) \mu(d). \qquad (2.3.9)$$

From $\log n = \sum_{d/n} \Lambda_d$ Lemma 2.3.2, it follows that

$$\sum_{n \le x} \log n = \sum_{n \le x} \sum_{kd=n} \Lambda_d = \sum_{k \le x} \sum_{d \le x/k} \Lambda_d = \sum_{k \le x} \psi\left(\frac{x}{k}\right).$$

Therefore, we obtain

$$\chi(x) = \sum_{n \le x} \log n - x \left[ \log x + C + O\left(\frac{1}{x}\right) \right] + [x] \log x - \sum_{k \le x} \log k + [x]C,$$

i.e.,

$$\chi(x) = O(\log(x+1)). \qquad (2.3.10)$$

This completes the proof of Lemma 2.3.10.                                                    □

**Lemma 2.3.11 ([Axer's Theorem] (Peyerimhoff [80, p. 87])).** *If*

(a) *$\chi(x)$ is of bounded variation in every finite interval $[1, T]$,*
(b) *$\sum_{1 \le k \le x} a_k = O(x)$,*
(c) *$a_n = O(1)$,*
(d) *$\chi(x) = O(x^\alpha)$ for some $0 < \alpha < 1$,*

*then*

$$\sum_{1 \le k \le x} \chi\left(\frac{x}{k}\right) a_k = O(x).$$

*Proof.* Let $0 < \delta < 1$. Then

$$\sum_{1 \le k \le \delta x} \chi\left(\frac{x}{k}\right) a_k = O(x^\alpha) \delta^{1-\alpha} x^{1-\alpha} = O\left(x\delta^{1-\alpha}\right).$$

Assuming that $m - 1 < \delta x \leq m$, $N \leq x < N + 1$ ($m$ and $N$ integers), we have

$$\sum_{\delta x \leq k \leq x} \chi\left(\frac{x}{k}\right) a_k = \sum_{k=m}^{N-1} \left[\chi\left(\frac{x}{k}\right) - \chi\left(\frac{x}{k+1}\right)\right] s_k + \chi\left(\frac{x}{N}\right) s_N - \chi\left(\frac{x}{m}\right) s_{m-1}$$

$$= O(x) \int_{\delta x}^{x} \left|d\chi\left(\frac{x}{t}\right)\right| + O(x)$$

$$= O(x) \int_{1}^{1/\delta} |d\chi(t)| + O(x).$$

This completes the proof of Lemma 2.3.11. □

**Lemma 2.3.12 (Peyerimhoff [80, p. 87]).** $\psi(x) - x = O(x)$.

*Proof.* It follows from (2.3.6), (2.3.9), (2.3.10), and Axer's theorem, that $\psi(x) - x = O(x)$.

This completes the proof of Lemma 2.3.12. □

**Lemma 2.3.13 (Peyerimhoff [80, p. 87]).** *Let* $\vartheta(x) = \sum_{p \leq x} \log p$ *(p prime),* *then*

(a) $\vartheta(x) \leq \psi(x) = O(x)$;
(b) $\psi(x) = \vartheta(x) + \vartheta(\sqrt{x}) + \cdots + \vartheta(\sqrt{x})$, *for every* $k > \frac{\log x}{\log 2}$.

**Lemma 2.3.14 (Peyerimhoff [80, p. 87]).**

$$\psi(x) = \vartheta(x) + O(1)\frac{\log x}{\log 2}\sqrt{x}.$$

*Proof.* It follows from part (b) of Lemma 2.3.13 that

$$\psi(x) = \vartheta(x) + O(1)\frac{\log x}{\log 2}\sqrt{x}.$$

This completes the proof of Lemma 2.3.14. □

**Lemma 2.3.15 (Peyerimhoff [80, p. 87]).**

$$\vartheta(x) = x + O(x). \tag{2.3.11}$$

*Proof.* Lemma 2.3.14 implies that $\vartheta(x) = x + O(x)$. □

## 2.4  The Prime Number Theorem

**Theorem 2.4.1.** *The PNT states that $\pi(x)$ is asymptotic to $x/\log x$ (see Hardy [41, p. 9]), that is, the limit of the quotient of the two functions $\pi(x)$ and $x/\ln x$ approaches 1, as $x$ becomes indefinitely large, which is the same thing as $[\pi(x)\log x]/x \to 1$, as $x \to \infty$ (Peyerimhoff [80, p. 88]).*

*Proof.* By definition and by

$$\pi(x) = \int_{3/2}^{x} \frac{1}{\log t} d\vartheta(t)$$

$$= \frac{\vartheta(x)}{\log x} + \int_{3/2}^{x} \frac{\vartheta(t)}{t(\log t)^2} dt$$

$$= \frac{\vartheta(x)}{\log x} + O\left(\frac{x}{(\log x)^2}\right)$$

[note that $\vartheta(x) = O(x)$].

Using (2.3.11) we obtain the PNT, i.e.,

$$\lim_{x\to\infty} \pi(x) = \frac{x}{\log x},$$

or

$$\lim_{x\to\infty} \frac{\pi(x)\log x}{x} = 1.$$

This completes the proof of Theorem 2.4.1.                                    □

# Chapter 3
# Summability Tests for Singular Points

## 3.1 Introduction

A point at which the function $f(z)$ ceases to be analytic, but in every neighborhood of which there are points of analyticity is called singular point of $f(z)$.

Consider a function $f(z)$ defined by the power series

$$f(z) = \sum_{n=0}^{\infty} a_n z^n \qquad (3.1.1)$$

having a positive radius of convergence. Every power series has a circle of convergence within which it converges and outside of which it diverges. The radius of this circle may be infinite, including the whole plane, or finite. For the purposes here, only a finite radius of convergence will be considered. Since the circle of convergence of the series passes through the singular point of the function which is nearest to the origin, the modulus of that singular point can be determined from the sequence $a_n$ in a simple manner. The problem of determining the exact position of the singular point on the circle of convergence is considered; tests can be devised to determine whether or not that point is a singular point of the function defined by the series. It may be supposed, without loss of generality, that the radius of convergence of the series is 1. In this chapter we apply Karamata/Euler summability method to determine or test if a particular point on the circle of convergence is a singular point of the function defined by the series (3.1.1).

## 3.2 Definitions and Notations

*Karamata's summability method* $K[\alpha, \beta]$ was introduced by Karamata (see [8]) and the summability method associated with this matrix is called Karamata method or $K[\alpha, \beta]$-method (c.f. [86]).

M. Mursaleen, *Applied Summability Methods*, SpringerBriefs in Mathematics,
DOI 10.1007/978-3-319-04609-9_3, © M. Mursaleen 2014

The *Karamata matrix* $K[\alpha, \beta] = (c_{nk})$ is defined by

$$c_{nk} = \begin{cases} 1, & n = k = 0, \\ 0, & n = 0, k = 1, 2, 3, \ldots, \end{cases}$$

$$\left[ \frac{\alpha + (1 - \alpha - \beta)z}{1 - \beta z} \right]^n = \sum_{k=0}^{\infty} c_{nk} z^k, \quad n = 1, 2, \ldots.$$

$K[\alpha, \beta]$ is the Euler matrix for $K[1 - r, 0] = E(r)$ (see [2]); the Laurent matrix for $K[1 - r, r] = S(r)$ (see [95]), and with a slight change, the Taylor matrix for $K[0, r] = T(r)$ (see [28]). If $T(r) = (c_{nk})$, then

$$\left[ \frac{(1 - r)z}{1 - rz} \right]^{n+1} = \sum_{k=0}^{\infty} c_{nk} z^{k+1}, \quad n = 0, 1, 2, \ldots$$

## 3.3   Tests for Singular Points

King [49] devised two tests in the form of following theorems, each of which provides necessary and sufficient condition that $z = 1$ be a singular point of the function defined by the series (3.1.1).

**Theorem 3.3.1.** *A necessary and sufficient condition that $z = 1$ be a singular point of the function defined by the series (3.1.1) is that*

$$\limsup_{n \to \infty} \left| \sum_{m=0}^{n} \binom{n}{m} r^m (1 - r)^{n-m} a_m \right|^{1/n} = 1,$$

*for some $0 < r < 1$.*

*Proof.* Consider the function

$$F(t) = \frac{1}{1 - (1 - r)t} f\left( \frac{rt}{1 - (1 - r)t} \right).$$

$F(t)$ is regular in the region

$$D_r = \left\{ t : \left| \frac{rt}{1 - (1 - r)t} \right| < 1 \right\}.$$

Furthermore, $z = 1$ is a singular point of $f(z)$ if and only if $t = 1$ is a singular point of $F(t)$. A simple calculation gives

$$D_r = \{t : \mathrm{Re}(t) < 1\},$$

$$D_r = \left\{ t : \left| t - \frac{1-r}{1-2r} \right| > \frac{r}{1-2r} \right\},$$

$$D_r = \left\{ t : \left| t - \frac{1-r}{1-2r} \right| < \frac{r}{2r-1} \right\},$$

for $r = 1/2$, $0 < r < 1/2$, and $1/2 < r < 1$, respectively. In each case $t = 1$ is on the boundary of $D_r$ and $D_r$ contains all points of the closed unit disk except $t = 1$. If we write $F(t) = \sum_{n=0}^{\infty} b_n t^n$, it follows that $t = 1$ is a singular point of $F(t)$ if and only if the radius of convergence of the series is exactly 1. That is, if and only if

$$\limsup_{n \to \infty} |b_n|^{1/n} = 1.$$

The function $F(t)$ is given by

$$F(t) = \frac{1}{1-(1-r)t} \sum_{m=0}^{\infty} a_m \left[ \frac{rt}{1-(1-r)t} \right]^m$$

$$= \sum_{m=0}^{\infty} a_m r^m t^m \sum_{n=m}^{\infty} \binom{n}{m} (1-r)^{n-m} t^{n-m}$$

provided that $(1-r)|t| < 1$. It is easy to verify the interchange of summation in the last expression. Hence, $F(t) = \sum_{n=0}^{\infty} t^n \sum_{m=0}^{n} \binom{n}{m} r^m (1-r)^{n-m} a_m$. Therefore,

$$b_n = \sum_{m=0}^{n} \binom{n}{m} r^m (1-r)^{n-m} a_m. \tag{3.3.1}$$

This completes the proof. $\qquad\square$

**Theorem 3.3.2.** *A necessary and sufficient condition that $z = 1$ be a singular point of the function defined by the series (3.1.1) is that*

$$\limsup_{m \to \infty} \left| \sum_{n=m}^{\infty} \binom{n}{m} r^{n-m} (1-r)^{m+1} a_n \right|^{1/n} = 1,$$

*for some $0 < r < 1$.*

*Proof.* Consider the function

$$G(t) = (1-r) \, f(r + (1-r)t).$$

$G(t)$ is regular in the region $R_r = \{t : |r + (1 - r)t| < 1\}$. A simple calculation gives

$$R_r = \left\{ t : \left| t - \frac{r}{r-1} \right| < \frac{1}{1-r} \right\}.$$

The point $t = 1$ is on the boundary of $R_r$ and $R_r$ contains all points of the closed unit disk except $t = 1$. If we write

$$G(t) = \sum_{n=0}^{\infty} c_n t^n,$$

it follows that $z = 1$ is a singular point of $f(z)$ if and only if

$$\limsup_{n \to \infty} |c_n|^{1/n} = 1.$$

The function $G(t)$ is given by

$$G(t) = (1 - r) \sum_{n=0}^{\infty} a_n (r + (1 - r)t)^n$$

$$= (1 - r) \sum_{n=0}^{\infty} a_n \sum_{m=0}^{\infty} \binom{n}{m} r^{n-m} (1 - r)^m t^m$$

$$= \sum_{m=0}^{\infty} t^m \sum_{n=m}^{\infty} \binom{n}{m} r^{n-m} (1 - r)^{m+1} a_n.$$

Hence,

$$c_m = \sum_{n=m}^{\infty} \binom{n}{m} r^{n-m} (1 - r)^{m+1} a_n. \tag{3.3.2}$$

This completes the proof.                                                                    □

These theorems yield the following corollaries.

**Corollary 3.3.3.** *If the sequence $(a_n)$ is $E(r)$-summable, $0 < r < 1$, to a nonzero constant, then $z = 1$ is a singular point of the function defined by the series (3.1.1).*

**Corollary 3.3.4.** *If the sequence $(a_n)$ is $T(r)$-summable, $0 < r < 1$, to a nonzero constant, then $z = 1$ is a singular point of the function defined by the series (3.1.1).*

Extending the above results, Hartmann [44] proved Theorem 3.3.6. The following lemma is needed for the proof of Theorem 3.3.6.

**Lemma 3.3.5.** *If* $K[\alpha, \beta] = (c_{nk})$ *for* $|\alpha| < 1, |\beta| < 1$, *then there exists* $\rho > 0$, *independent of* $k$, *such that for* $|t| < \rho$ *and* $k = 0, 1, 2, \ldots$,

$$\sum_{n=0}^{\infty} c_{n,k+1} t^n = \frac{(1-\alpha)(1-\beta)t}{(1-\alpha t)^2} \left[ \frac{\beta + (1-\alpha-\beta)t}{1-\alpha t} \right]^k.$$

*Proof.* Let $f(z) = [\alpha + (1-\alpha-\beta)z]/(1-\beta z)$. If $0 < R < 1 < 1/|\beta|$, then there exists $\rho_1 > 0$ such that if $|t| \leq \rho_1$ and let

$$\phi_t(z) = \frac{1}{1-tf(z)} = \sum_{n=0}^{\infty} t^n [f(z)].$$

Since this convergence is uniform in $|z| \leq R$, one can apply Weierstrass theorem on uniformly convergent series of analytic functions (see [53]) to write

$$\sum_{n=0}^{\infty} t^n [f(z)]^n = \sum_{n=0}^{\infty} t^n \left( \sum_{k=0}^{\infty} c_{nk} z^k \right) = \sum_{k=0}^{\infty} z^k \left( \sum_{n=0}^{\infty} c_{nk} t^n \right). \qquad (3.3.3)$$

But

$$\frac{1}{1-tf(z)} = \frac{1-\beta z}{(1-\alpha t)\left[ 1 - \frac{\beta+(1-\alpha-\beta)t}{1-\alpha t} z \right]}. \qquad (3.3.4)$$

There exits $\rho_2 > 0$ such that $|t| \leq \rho_2$ and $|z| \leq R$ imply $|[\beta + (1-\alpha-\beta)t]z/[1-\alpha t]| < 1$. Thus (3.3.4) may be expanded in a power series,

$$\frac{1}{1-tf(z)} = \sum_{k=0}^{\infty} \frac{1-\beta z}{1-\alpha t} \left[ \frac{\beta + (1-\alpha-\beta)t}{1-\alpha t} \right]^k z^k. \qquad (3.3.5)$$

Then, for $|t| \leq \min(\rho_1, \rho_2)$, one has, by equating coefficients in (3.3.3) and (3.3.5), the results of the lemma. □

**Theorem 3.3.6.** *A necessary and sufficient condition that* $z = 1$ *be a singular point of the function defined by the series (3.1.1) is that*

$$\limsup_{n \to \infty} \left| \sum_{k=0}^{\infty} c_{n,k+1} a_k \right|^{1/n} = 1 \qquad (3.3.6)$$

*for some* $\alpha < 1, \beta < 1$ *and* $\alpha + \beta > 0$.

*Proof.* Consider the function

$$F(t) = \frac{(1-\alpha)(1-\beta)t}{(1-\alpha t)^2} f\left(\frac{\beta + (1-\alpha-\beta)t}{1-\alpha t}\right).$$

$F(t)$ is regular in the region $D$, where

$$D = \left\{t : \left|\frac{\beta + (1-\alpha-\beta)t}{1-\alpha t}\right| < 1\right\}.$$

Furthermore, $z = 1$ is a singular point of $f(z)$ if and only if $t = 1$ is a singular point of $F(t)$. A simple calculation gives

$$D = \begin{cases} t : |t + \frac{\alpha+\beta}{1-\beta-2\alpha}| < |\frac{1-\alpha}{1-\beta-2\alpha}| , 1-\beta-2\alpha > 0; \\ t : \text{Re}\,(t) < 1 \quad\quad , 1-\beta-2\alpha = 0; \\ t : |t + \frac{\alpha+\beta}{1-\beta-2\alpha}| > |\frac{1-\alpha}{1-\beta-2\alpha}| , 1-\beta-2\alpha < 0. \end{cases}$$

In each case $t = 1$ is on the boundary of $D$ and $D$ contains all points of the closed unit disk except $t = 1$. Writing $F(t)$ in series form yields

$$F(t) = \frac{(1-\alpha)(1-\beta)t}{(1-\alpha t)^2} \sum_{k=0}^{\infty} a_k \left[\frac{\beta + (1-\alpha-\beta)t}{1-\alpha t}\right]^k,$$

provided $t \,\varepsilon\, D$. By Lemma 3.3.5, there exists $\rho > 0$ such that for $|t| \le \rho_1 < \rho$ and $k = 0, 1, 2, \ldots$

$$\sum_{n=0}^{\infty} c_{n,k+1} t^n = \frac{(1-\alpha)(1-\beta)t}{(1-\alpha t)^2} \left[\frac{\beta + (1-\alpha-\beta)t}{1-\alpha t}\right]^k. \tag{3.3.7}$$

Since $(1-\alpha)(1-\beta)t/(1-\alpha t)^2$ vanishes for $t = 0$ and $[\beta + (1-\alpha-\beta)t]/[1-\alpha t]$ is equal to $\beta$ for $t = 0$, with $|\beta| < 1$, there exists $\rho_2(\alpha,\beta) < \rho_1$ such that $|t| \le \rho_2$ implies $|\sum_{n=0}^{\infty} c_{n,k+1} t^n| \le M r^k$ for some $r = r(\alpha,\beta) < 1$. Thus

$$\left|\sum_{k=0}^{\infty}\sum_{n=0}^{\infty} c_{n,k+1} a_k t^n\right| \le \sum_{k=0}^{\infty} |a_k| \left|\sum_{n=0}^{\infty} c_{n,k+1} t^n\right|$$

$$\le M \sum_{k=0}^{\infty} |a_k| r^k,$$

which converges since (3.3.7) has radius of convergence one. Weierstrass theorem now implies

$$F(t) = \sum_{k=0}^{\infty}\sum_{n=0}^{\infty} c_{n,k+1} a_k t^n, \tag{3.3.8}$$

for $|t| \leq \rho_2$. By analytic continuation (3.3.8) holds in a disk whose boundary contains the singularity of $F(t)$ nearest the origin and $t = 1$ is a singular point of $F(t)$ if and only if the radius of convergence of series (3.3.8) is exactly 1, i.e.,

$$\limsup_{n \to \infty} \left| \sum_{k=0}^{\infty} c_{n,k+1} a_k \right|^{1/n} = 1. \tag{3.3.9}$$

This completes the proof of the theorem.                                              $\square$

From this, following result may be deduced.

**Corollary 3.3.7.** *If the sequence $(0, a_0, a_1, \ldots)$ is $K[\alpha, \beta]$ summable $\alpha < 1, \beta < 1, \alpha + \beta > 0$, to a nonzero constant, then $z = 1$ is a singular point of the function given by (3.1.1).*

*Remark 3.3.8.* Notice $K[\alpha, \beta]$ is regular for $\alpha < 1, \beta < 1$ and $\alpha + \beta > 0$ (see [8]). If $(b_n)$ is the $K[\alpha, \beta]$ transform of $(0, a_0, a_1, \ldots)$, then $b_0 = 0$, $b_n = \sum_{k=0}^{\infty} c_{n,k+1} a_k$, $n = 1, 2, \ldots$. Now, if $(0, a_0, a_1, \ldots)$ is $K[\alpha, \beta]$ summable to a nonzero constant, then (3.3.6) holds. If the $T(r)$ transform of $(a_n)$ is $(c_n)$ and the $K[0, r]$ transform of $(0, a_0, a_1, \ldots)$ is $(\gamma_n)$, then $\gamma_0 = 0$, $\gamma_n = c_{n-1} (n \geq 1)$ and thus one has immediately Corollary 3.3.4. In [2] it is proved that $E(r)$ is translative to the right when $E(r)$ is regular, so Corollary 3.3.7 implies Corollary 3.3.3.

# Chapter 4
# Lototski Summability and Analytic Continuation

## 4.1 Introduction

Analytic continuation is a technique to extend the domain of a given analytic function. Analytic continuation often succeeds in defining further values of a function, for example, in a new region where an infinite series representation in terms of which it is initially defined becomes divergent.

The problem of analytic continuation by summability may be formulated as follows: Let $f(z)$ have the Taylor expansion

$$f(z) = \sum_{k=0}^{n} a_k (z - z_0)^k \qquad (4.1.1)$$

with a positive radius of convergence. Two questions arise: (i) What is the condition of efficiency of a special linear transformation of (4.1.1) regarding the analytic continuation of $f(z)$? (ii) Given some domain in the complex plane, does there exist a linear transformation of (4.1.1) which yields the analytic continuation of $f(z)$ exactly into this domain and nowhere else? In some cases, as has been shown by Borel [18], Okada [78], and Vermes [96], it is sufficient to focus attention on the continuation of the geometric series $\sum z^n$, $|z| < 1$; in this chapter we deal only with the series in (4.1.1). In this context, Cooke and Dienes [27] have shown that there exist transformations that are effective at some distinct points outside the circle of convergence; this result was extended by Vermes [97] to a denumerable set of points. Russel [85] and Teghem [94] have produced transformations effective respectively on Jordan arcs and on domains that are not simply connected. In this chapter, we describe a new Toeplitz summability method, i.e., the generalized Lototski or $[F, d_n]$-summability, and study the regions in which these methods sum a Taylor series to the analytic continuation of the function which it represents.

M. Mursaleen, *Applied Summability Methods*, SpringerBriefs in Mathematics, DOI 10.1007/978-3-319-04609-9_4, © M. Mursaleen 2014

## 4.2   Definitions and Notations

(i) Suppose $f$ is an analytic function defined on an open subset $U$ of the complex
plane. If $V$ is a larger open subset of the complex plane containing $U$ and $g$
is an analytic function defined on $V$ such that

$$g(z) = f(z) \text{ for all } z \in U,$$

then $g$ is called an *analytic continuation of $f$*. In other words, the restriction
of $g$ to $U$ is the function $f$ we started with.

(ii) Corresponding to a real or complex sequence $(d_k)$ such that $d_k \neq -1$ for all
$k \in \mathbb{N}$, the *generalized Lototski* or *$[F, d_n]$-transform* $(t_n)$ of a sequence $(s_n)$
is defined by (Jakimovski [46])

$$t_n = \prod_{k=1}^{n} \frac{(d_k + E)(s_0)}{d_k + 1}, \quad n \geq 1, \tag{4.2.1}$$

where

$$E^p(s_k) = s_{p+k}, \quad k \geq 0, \ p \geq 0.$$

If $\lim t_n$ exists as $n \to \infty$, we say that $(s_n)$ is summable $[F, d_n]$ to the value
$\lim t_n$.

(iii) For every sequence of polynomials $\{P_n(x)\}$ satisfying $P_n(1) \neq 0$, the
$[F^*, P_n]$-transform of a sequence $(s_n)$ will be defined by

$$t_n^* = \prod_{k=0}^{n} \frac{P_k(E)(s_0)}{P_k(1)}, \quad n \geq 1. \tag{4.2.2}$$

It may easily be seen that if $(s_n)$ is the sequence of sums of the geometric
series $\sum_{n=0}^{\infty} z^n$, $(z \neq 1)$; then in the notation above

$$t_n = \frac{1}{1-z} - \frac{z}{1-z} \prod_{k=1}^{n} \frac{d_k + z}{d_k + 1} \tag{4.2.3}$$

and

$$t_n^* = \frac{1}{1-z} - \frac{z}{1-z} \prod_{k=1}^{n} \frac{P_k(z)}{P_k(1)}. \tag{4.2.4}$$

It follows that $t_n \to (1 - z)^{-1}$, as $n \to \infty$ if and only if

$$\lim_{n\to\infty} \prod \frac{d_k + z}{d_k + 1} = 0, \tag{4.2.5}$$

while $t_n^* \to 1/(1 - z)$, as $n \to \infty$ if and only if

$$\lim_{n\to\infty} \prod_{k=1}^{n} \frac{P_k(z)}{P_k(1)} = 0. \tag{4.2.6}$$

(iv) By $\Gamma$, we denote a family of Jordan arcs $\gamma$ in the complete complex plane, with end points $0$, $\infty$ directed from $0$ to $\infty$, and having the following properties: (a) If $\gamma_1$ and $\gamma_2$ are two different elements of $\Gamma$, then they intersect only at $0$ and $\infty$; (b) to each complex $z$ ($z \neq 0, \infty$) corresponds an element $\gamma(z) = \gamma \in \Gamma$ passing through $z$. We write $[0, z]$ and $[z, \infty]$ for the subarcs of $\gamma(z)$ with end points $0$ and $z$ and with end points $z$ and $\infty$, respectively, and we replace brackets by parenthesis to indicate the corresponding end point is deleted from the subarc.

(v) If $A$ and $B$ are two point sets, we denote:

(a) by $d(A, B)$, the distance between them;
(b) by $A^{-1}$, the set $\{z : z^{-1} \in A\}$;
(c) by $AB$, the set $\{s : s = zw, z \in A, w \in B\}$;
(d) by $wA$, the set $\{s : s = zw, z \in A\}$;
(e) by $A^c$, the complement of $A$ relative to the complete complex plane.

(vi) A family $\Gamma$ will be called continuous provided to each $z \neq 0, \infty$, and each $\varepsilon > 0$ there corresponds a $\delta = \delta(z_1, \varepsilon) > 0$, such that

$$\sup_{w\in[0,z]} d(w, [0, z_1]) < \varepsilon,$$

for all points $z$ in the disk $|z - z_1| < \delta$. The following example shows that an arbitrary family $\gamma$ is not necessarily continuous. Let $\gamma_0$ be the linear ray $z \geq 0$. For $n \geq 1$, let $\gamma_n$ be the polygonal line composed of the two line segments $[0, 3 + 3i/2^n]$ and $[3 + 3i/2^{2n}, 2 + 3i/2^{2n+1}]$ and the ray $t + 3i/2^{2n+1}$, ($t \geq 2$). We can easily embed the sequence $(\gamma_n)_0^\infty$ in a family $\gamma$ (not uniquely). Suppose this is done, and choose $z_1 = 2$ and $z = 2 + 2^{-(n+1)}i$. Then,

$$\sup_{w\in[0,z]} d(w, [0, 2]) \geq d(3 + 3i/2^{2n}, 2) > 1.$$

Choose $\varepsilon = 1$, we see that $\gamma$ is not continuous.

(vii) Denote by $P \equiv P(z)$, a power series $\sum_{n=0}^\infty a_n z^n$ with the partial sums $s_n(z)$ and with a positive radius of convergence. Continue $P(z)$ analytically along each $\gamma \in \Gamma$ from $0$ to the first singular point $w(\gamma)$ on $\gamma$. If there is no finite singular point on $\gamma$, we define $w(\gamma) = \infty$. By $M \equiv M(P; \Gamma)$ we denote the union of all the sets $[0, w(\gamma))$, and we call this set the $\Gamma$-*Mittag-Leffler star* of $P(z)$. Clearly, $\infty \notin M$ (se [47]). If $z_0 \in M$, we denote by $P(z_0; \Gamma)$, the value at $z_0$ of the analytic continuation of $P(z)$ along $\gamma(z_0)$. By definition, $P(z; \Gamma)$ is a single valued function in $M$. A set $D$ is called a $\Gamma$-*star* set provided it is not empty, $\infty \notin D$, and $z \in D$ implies $[0, z] \subset D$. A $\Gamma$-star set that is also a domain. Obviously, a $\Gamma$-star domain is simply connected, a union of $\Gamma$-star domains, and an intersection of $\Gamma$-star sets is a $\Gamma$-star set.

(viii) For a family $\Gamma$, we define the set $D(\Gamma)$ by

$$D(\Gamma) = \left\{ s = \frac{z}{w}, \ z \neq 0, \infty; w \in (0, z] \right\}.$$

A set D is $\Gamma$-*regular* if $0 \in D$, $1 \notin D$, $\infty \notin D$, and $D(\Gamma) \subset D^c$.


## 4.3  Main Results

We discuss in this chapter the results obtained by Meir [5] and Jakimovski [47]. Generalizing some known results Meir [5] proved the following theorems.

**Theorem 4.3.1.** *Let the polynomial $P(z)$ satisfy*

$$\mathrm{Re}[P(1)] = 0. \tag{4.3.1}$$

*Then, there exists a fixed sequence $(d_n)$, $(n \geq 1$ and $d_n \neq -1)$ such that $[F, d_n]$-transform sums the geometric series to the value $(1 - z)^{-1}$ for every $z$ for which $\mathrm{Re}[P(z)] > 0$ and does not sum it for every $z$ for which $\mathrm{Re}[P(z)] < 0$. The convergence of the transform is uniform in every bounded closed subset of $\{z, \mathrm{Re}[P(z)] > 0\}$.*

*Proof.* Clearly we may suppose $P(z) \neq$ constant. Then for every $k \geq 1$

$$P(z) + k = c \left( z + a_1^k \right) \left( z + a_2^k \right) \cdots \left( z + a_p^k \right), \tag{4.3.2}$$

where $p \geq 1$, $c \neq 0$ and $c$ does not depend on $k$. Define now $d_1 = a_1'$, $d_2 = a_2', \ldots, d_p = a_1^2, \ldots, d_{2p}, \ldots$ and in general if $v = \mu p + \rho$, $(0 < \rho \leq p)$

$$d_v = a_p^{\mu+1}. \tag{4.3.3}$$

Now let $n = mp + q$, $(0 \leq q < p)$; then

$$\prod_{v=1}^{n} \frac{d_v + z}{d_v + 1} = \prod_{k=1}^{n} \frac{P(z) + k}{P(1) + k} \left( \prod_{v=mp+1}^{mp+q} \frac{d_v + z}{d_v + 1} \right) = \Pi_1^{(n)} \Pi_2^{(n)}, \tag{4.3.4}$$

where the second factor is 1 if $q = 0$. By (4.3.1), if $|1-z| < \delta$, then $|\operatorname{Re} P(z)| < 1/2$, and by (4.3.2) and (4.3.3) for $1 \le \rho \le p,\ \mu \ge 0$,

$$P(-d_{\mu p+\rho}) = -(\mu + 1) \le -1; \tag{4.3.5}$$

thus

$$|1 + d_\nu| \ge \delta > 0, \quad \nu = 1, 2, \ldots \tag{4.3.6}$$

$$\left| \Pi_2^{(n)} \right| = \left| \prod_{\nu=np+1}^{np+q} \left( 1 + \frac{z-1}{d_\nu + 1} \right) \right| \le \prod_{\nu=np+1}^{np+q} \left( 1 + \left| \frac{z-1}{d_\nu + 1} \right| \right), \tag{4.3.7}$$

and by (4.3.6)

$$\left| \Pi_2^{(n)} \right| \le \left( 1 + \left( \frac{|z-1|}{\delta} \right) \right)^{p-1}.$$

Thus $\Pi_2^{(n)}$ is uniformly bounded for every $n \ge 1$ and for every $z$ belonging to a fixed bounded point set.

$$\left| \Pi_1^{(n)} \right|^2 = \prod_{k=1}^{n} \left| \frac{P(z) + k}{P(1) + k} \right|^2$$

$$= \prod_{k=1}^{n} \left\{ 1 + \frac{2k\operatorname{Re}[P(z)] + |P(z)|^2 - |P(1)|^2}{k^2 + |P(1)|^2} \right\}. \tag{4.3.8}$$

By a well-known theorem on infinite products

$$\lim_{n\to\infty} \Pi_1^{(n)} = \begin{cases} 0 , & \operatorname{Re}[P(z)] < 0, \\ \infty , & \operatorname{Re}[P(z)] > 0. \end{cases} \tag{4.3.9}$$

Also, the convergence to 0 is uniform in every point set where $\operatorname{Re} P(z) \le -\varepsilon$, with $\varepsilon > 0$ fixed. (4.3.9), (4.3.7), (4.3.4), and (4.3.5) prove the theorem. $\qquad\square$

**Theorem 4.3.2.** *Let $R$ be a bounded set that contains the point $z = 1$ and whose complement consists either of the point $\infty$ or of an unbounded domain. Let $f(z)$ be an analytic regular function satisfying*

$$\operatorname{Re}[f(1)] = 0. \tag{4.3.10}$$

*Then, there exists a sequence of polynomials $\{P_n(x)\}$, $(n \ge 1,\ P_n(1) \ne 0)$ such that $[F^*, P_n]$-transformation sums the geometric series to the value $(1 - z)^{-1}$ for every $z \in R$ for which $\operatorname{Re}[f(z)] < 0$ and does not sum it for $z \in R$ for which $\operatorname{Re}[f(z)] > 0$.*

*Proof.* By the well-known theorem of Walsh [98], for every $k \geq 1$ there exist polynomials $Q_k(z)$ satisfying

$$|Q_k(z) - f(z)| < \frac{1}{k} \qquad (4.3.11)$$

for $z \in R$ with $|z| \leq k$, and

$$Q_k(1) = f(1); \quad k = 1, 2, \ldots. \qquad (4.3.12)$$

Define

$$P_k(z) = Q_k(z) + k; \quad k = 1, 2, \ldots. \qquad (4.3.13)$$

By (4.3.11)–(4.3.13) for any fixed $z$ such that $|z| \leq K$

$$\frac{P_k(z)}{P_k(1)} = 1 + \frac{f(z) - f(1)}{k} + O\left(\frac{1}{k^2}\right).$$

Now, by (4.3.2) and the theory of infinite products, if $z \in R$

$$\lim_{n \to \infty} \prod_{k=1}^{n} \frac{P_k(z)}{P_k(1)} = \begin{cases} 0 \ , \ \mathrm{Re}[P(z)] < 0, \\ \infty \ , \ \mathrm{Re}[P(z)] > 0, \end{cases}$$

by (4.2.6).

This completes the proof of the theorem.                                               □

*Remark 4.3.3.* A generalization of Theorem 4.3.2 can be made to the situation where R is the union of increasing sequence of bounded closed sets $R_i$ the complement of each of which is an unbounded domain. This result will prove the existence of an $[F^*, P_n]$-transformation that is effective for $\sum z^n$ in the Mittag-Leffler star of $(1 - z)^{-1}$. It has to be mentioned that the $[F^*, P_n]$-transformations are row-finite. Because of the lengthy proof Meir only stated the following result too:

**Theorem 4.3.4.** *Let D be an union of a finite number of simply connected bounded domains having Jordan boundaries. Let $z = 1$ lie on the boundary, and let E be a closed subset of the complement of D. Then, there exists an $[F^*, P_n]$-transformation, which sums the geometric series to the sum $(1 - z)^{-1}$ for every $z \in D$ and does not sum it for every $z \in E$.*

In the more generalized setup, Jakimovski [47] proved the following:

**Theorem 4.3.5.** *Let $\Gamma$ be continuous. Suppose the infinite matrix $(a_{nm})_{n,m=0}^{\infty}$ has the following properties:*

(i) $\sum_{m=0}^{\infty} a_{nm} \to 1$, *as $n \to \infty$.*

*(ii) For certain open and $\Gamma$-regular set D, the relation*

$$\lim_{n \to \infty} \sum_{m=0}^{\infty} a_{nm} z^{m+1} = 0$$

*holds uniformly in every compact subset of D. Then, for each power series $P(z)$ with a positive radius of convergence, the relation*

$$\lim_{n \to \infty} \sum_{m=0}^{\infty} a_{nm} \, s_m(z) = P(z, \Gamma) \tag{4.3.14}$$

*holds uniformly in each compact subset of the set $\Omega = \cap \{wD : w \notin M, \, w \notin \infty\}$, where M is defined in part (vii).*

The following lemmas are needed for the proof of Theorem 4.3.5.

**Lemma 4.3.6.** *If $\Gamma$ is continuous, then $M(P; \Gamma)$ is a simply connected domain and $P(z; \Gamma)$ is holomorphic in $M(P; \Gamma)$. If $\Gamma$ is not continuous, then $M(P; \Gamma)$ is not necessarily a domain.*

*Proof.* We have to show that if $\Gamma$ is continuous, then $M(P; \Gamma)$ is a simply connected domain and $\frac{d}{dz} P(z; \Gamma)$ exists for all $z \in M$. If $z_0 \in M$ and $z_0 \neq 0$, then there exists a domain $G$ and a function $f$, holomorphic in $G$, such that $[0, z_0] \subset G$ and $f(z) = P(z; \Gamma)$ for $z \in [0, z_0]$. The continuity of $\Gamma$ implies the existence of a $\delta > 0$ such that $[0, z] \subset G$ whenever $|z - z_0| < \delta$. Therefore

$$\{z : |z - z_0| < \delta\} \subset M(P; \Gamma)$$

and $P(z; \Gamma) = f(z)$ for these values of $z$. Thus $P'(z_0; \Gamma)$ exists and $M(P; \Gamma)$ is an open set. The first part of lemma now follows from the general properties of $\Gamma$-star sets.

Next, let $\Gamma$ be the noncontinuous family described earlier. In order to prove Lemma 4.3.6 it is enough to show the existence of the power series $P(z)$ with a positive radius of convergence such that $M(P; \Gamma)$ is not a domain. Choose

$$a_n = \frac{5}{2} + \frac{19i}{2^{2n+3}}, \quad b_n = 2 + \frac{3i}{2^{2n+1}}; \quad (n \geq 1).$$

For the function $\log\{(z-a_n)/(b_n-a_n)\}(n \geq 1)$, choose at $z = 0$ the branch which, if continued analytically from 0 to $b_n$ along the linear segment $[0, b_n]$, yields at $z = b_n$ the value $\log 1 = 2\pi i$. The function

$$P(z) = \sum_{n=1}^{\infty} \frac{1}{n! \log \frac{z-a_n}{z_n-b_n}}$$

is holomorphic in $|z| < 5/2$. For the Jordan arc $\gamma_0$ of our discontinuous family, $w(\gamma_0) = 5/2$. For the Jordan arcs $\gamma_n'$, $w(\gamma_n) = b_n$. This means that

$$\left\{ z : 0 \leq z < \frac{5}{2} \right\} \subset M(P;\Gamma)$$

and

$$\left\{ z : z = t + \frac{3i}{2^{2n+1}}, t \geq 2 \right\} \subset M(P;\Gamma)^C, \quad (n \geq 1).$$

Hence each point $z$ with $2 \leq z < 5/2$ is not an interior point of $M(P;\Gamma)$, and $M(P;\Gamma)$ is not a domain.

This completes the proof of Lemma 4.3.6. □

**Lemma 4.3.7.** *Let $D$ be a $\Gamma$-regular set. Suppose $\gamma$ is a bounded Jordan curve whose interior contains the point $0$. If a set $F$ satisfies the condition $F \subset \cap_{w \in \gamma} w D$, then it lies in the interior of $\gamma$.*

*Proof.* If $z$ is on $\gamma$ or in the exterior of $\gamma$, then $z \neq 0$ and there exists a point $z_1$ such that $z_1 \in (0, z]$, $z_1 \in \gamma$, and $[0, z_1)$ is included in the interior of $\gamma$. Hence $z z_1^{-1} \in D(\Gamma) \subset D^C$. The last fact and hypothesis on $F$ imply that $z \notin F$.

This completes the proof of Lemma 4.3.7 □

*Proof of Theorem 4.3.5.* Suppose that $F$ is any compact set in $\Omega$ and $0 \in F$. First we establish the existence of rectifiable Jordan curve $\gamma$ with the following three properties:

(a) $\gamma \subset M(P;\Gamma)$;
(b) $\Gamma \gamma^{-1} \subset D$;
(c) $F$ lies in the interior of $\gamma$.

Since $M(P;\Gamma)$ is a $\Gamma$-star set, Lemma 4.3.6 and our hypothesis on $F$ imply that

$$F(M^C)^{-1} \subset D \text{ and } \delta \equiv d(F(M^C)^{-1}, D^C) > 0.$$

Because the set $(M^C)^{-1}$ is a bounded continuum, there corresponds to each $a > 0$ a rectifiable Jordan curve $\xi$ that includes $(M^C)^{-1}$ in its interior and has the property

$$\sup_{w \in \xi} d(w, (M^C)^{-1}) < \frac{\delta}{4a}.$$

Let $\gamma = \xi^{-1}$. Then $\gamma$ obviously has property (a). Since $F$ is bounded (say $|z| \leq a$ for all $z \in F$), there corresponds to each $u \in \gamma$ a point $w = w(u) \in M^C$ such that $|u^{-1} - w^{-1}| < \delta/4a$, whence $|z/u - z/w| < \delta/4$ for all $z \in F$. Thus

$$d\left(\frac{z}{u}, D^C\right) \geq d\left(\frac{z}{w}, D^C\right) - \left|\frac{z}{u} - \frac{z}{w}\right| > \frac{\delta}{2}.$$

Therefore $d(F\gamma^{-1}, D^C) > \delta/2$. In particular, $\gamma$ has property (b). Since property (b) is equivalent to the assumption that $F \subset \Omega$, Lemma 4.3.7 implies that $\gamma$ has property (c). Lemma 4.3.6, the properties of $\gamma$, the fact that $1 \notin D$, the assumption (ii) of Theorem 4.3.5, and the calculus of residues yield the relation

$$P(z; \Gamma) = \frac{1}{2\pi i} \oint_\gamma \frac{P(w; \Gamma)}{w} \lim_{n \to \infty} \sum_{m=0}^{\infty} a_{nm} \frac{1 - (\frac{z}{w})^{m+1}}{1 - \frac{z}{w}} dw$$

$$= \lim_{n \to \infty} \sum_{m=0}^{\infty} a_{nm} s_m(z) \qquad (4.3.15)$$

for all $z \in F$, and the convergence is uniform in $F$.

This completes the proof of Theorem 4.3.5. $\qquad\qquad\square$

*Remark 4.3.8.* It is easy to see ([47], Remark, p. 355) that the assumptions of Theorem 4.3.5 imply that $\Omega \subset M (P; \Gamma)$ (so that the right hand of (4.3.14) is defined) and that the set of finite points of $\Omega$ is open.

*Example 4.3.9.* The following statements hold:

(i) The Lototski transform defined by $[F, d_n = n - 1]$ sums the geometric series for $\text{Re}(z) < 1$ and does not sum it for $\text{Re}(z) > 1$, [46]. Here $P(z) = z - 1$.

(ii) If $P(z) = e^{i\gamma}(z - 1)$ with a suitable real $\gamma$, we obtain a domain of summability any given half plane, the boundary of which is a straight line passing through $z = 1$.

(iii) The family $\Gamma$ of all rays emanating from the point 0 is continuous and has the property that $D(\Gamma) = \{x : x \geq 1\}$. In this special case, $M(p; \Gamma)$ is the ordinary Mittag-Leffler star of $P(z)$, and Theorem 4.3.5 is a generalization of Okada's theorem. Here we have the additional result about the uniform summability in compact subsets, which has proved for special domains $D$ in [78] (see also [26, p. 189]).

(iv) Let $\gamma$ be a Jordan arc defined by $\phi = \phi(r)$ for $z = re^{i\phi}$, where $\phi$ is continuous for $0 \leq r < \infty$. The family of all Jordan arcs of the form $\gamma_\alpha = e^{i\alpha}\gamma$; $(0 \leq \alpha < 2\pi)$ is continuous. In particular case where $\gamma$ is a polynomial line composed of the line segment $[0, 1]$ and the ray $1 - iy$, $(0 \leq y < +\infty)$,

$$D(\Gamma)^C = \{z : z \geq 1, z \leq 0\}.$$

# Chapter 5
# Summability Methods for Random Variables

## 5.1 Introduction

Let $(X_k)$ be a sequence of independent, identically distributed (i.i.d.) random variables with $E|X_k| < \infty$ and $EX_k = \mu$, $k = 1, 2, \ldots$. Let $A = (a_{nk})$ be a Toeplitz matrix, i.e., the conditions (1.3.1)–(1.3.3) of Theorem 1.3.3 are satisfied by the matrix $A = (a_{nk})$. Since

$$E \sum_{k=1}^{\infty} |a_{nk} X_k| = E|X_k| \sum_{k=1}^{\infty} |a_{nk}| \leq ME|X_k|,$$

the series $\sum_{k=0}^{\infty} a_{nk} X_k$ converges absolutely with probability one.

There is a vast literature on the application of summability to Probability Theory. Here, we study only few applications of summability methods in summing sequences of random variables and strong law of large numbers (c.f. [86]).

## 5.2 Definitions and Notations

In this section, we give some required definitions.

**Definition 5.2.1 (Random variables).** A function $X$ whose range is a set of real numbers, whose domain is the sample space (set of all possible outcomes) $S$ of an experiment, and for which the set of all $s$ in $S$, for which $X(s) \leq x$ is an event if $x$ is any real number. It is understood that a probability function is given that specifies the probability $X$ has certain values (or values in certain sets). In fact, one might define a random variable to be simply a probability function $P$ on suitable subsets of a set $T$, the point of $T$ being "elementary events" and each set in the domain of $P$ an event.

M. Mursaleen, *Applied Summability Methods*, SpringerBriefs in Mathematics,
DOI 10.1007/978-3-319-04609-9_5, © M. Mursaleen 2014

**Definition 5.2.2 (Independent random variables).** Random variables $X$ and $Y$ such that whenever $A$ and $B$ are events associated with $X$ and $Y$, respectively, the probability $P(A \text{ and } B)$ of both is equal to $P(A) \times P(B)$.

**Definition 5.2.3 (Distribution).** A random variable together with its probability density function, probability function, or distribution function is known as *distribution*.

**Definition 5.2.4 (Distribution function).** A real-valued function $G(x)$ on $R = [-\infty, \infty]$ is called *distribution function* (abbreviated d.f.) if $G$ has the following properties:

(a)  $G$ is nondecreasing;
(b)  $G$ is left continuous, i.e., $\lim_{y \to x, y < x} G(y) = G(x)$, all $x \in R$;
(c)  $G(-\infty) = \lim_{x \to -\infty} G(x) = 0$, $G(\infty) = \lim_{x \to \infty} G(x) = 1$.

**Definition 5.2.5 (Independent, identically distributed random variable).** A sequence $(X_n)_{n \geq 1}$ (or the random variables comprising this sequence) is called independent, identically distributed (abbreviated i.i.d.) if $X_n, n \geq 1$, are independent and their distribution functions are identical.

**Definition 5.2.6 ($\sigma$-field).** A class of sets $F$ satisfying the following conditions is called a $\sigma$-field:

(a)  if $E_i \in F$ ($i = 1, 2, 3, \ldots$), then $\cup_{i=1}^n E_i \in F$;
(b)  if $E \in F$, then $E^c \in F$.

**Definition 5.2.7 (Probability Space).** Let $F$ be a $\sigma$-field of subsets of $\Omega$, i.e., nonempty class of subsets of $\Omega$ which contains $\Omega$ and is closed under countable union and complementation. Let $P$ be a measure defined on $F$ satisfying $P(\Omega) = 1$. Then the triple $(\Omega, F, P)$ is called *probability space*.

**Definition 5.2.8 (Expectation).** Let $f$ be the relative frequency function (probability density function) of the variable $x$. Then

$$E(x) = \int_a^b x f(x) dx$$

is the expectation of variable $x$ over the range $a$ to $b$, or more usually, $-\infty$ to $\infty$.

**Definition 5.2.9 (Almost Everywhere).** A property of points $x$ is said to hold *almost everywhere*, a.e., or for almost all points, if it holds for all points except those of a set of measure zero.

The concept of almost sure (a.s.) convergence in probability theory is identical with the concept of almost everywhere (a.e.) convergence in measure theory.

**Definition 5.2.10 (Almost Sure).** The sequence of random variables $(X_n)$ is said to *converge almost sure*, in short a.s. to the random variable $X$ if and only if there exists a set $E \in F$ with $P(E) = 0$, such that, for every $w \in E^c$, $|X_n(w) - X(w)| \to 0$, as $n \to \infty$. In this case, we write $X_n \overset{a.s.}{\to} X$.

**Definition 5.2.11 (Median).** For any random variable $X$ a real number $m(X)$ is called a *median of X* if $P\{X \leq m(X)\} \geq (1/2) \leq P\{X \geq m(X)\}$.

**Definition 5.2.12 (Levy's inequalities).** If $\{X_j; \ 1 \leq j \leq n\}$ are independent random variables and if $S_j = \sum_{i=1}^{j} X_i$, and $m(Y)$ denotes a median of $Y$, then, for any $\epsilon > 0$,

(i) $P\{\max_{1 \leq j \leq n}[S_j - m(S_j - S_n)] \geq \epsilon\} \leq 2P\{|S_n| \geq \epsilon\}$;
(ii) $P\{\max_{1 \leq j \leq n}|S_j - m(S_j - S_n)| \geq \epsilon\} \leq 2P\{S_n \geq \epsilon\}$.

**Definition 5.2.13 (Chebyshev's inequality).** In probability theory, *Chebyshev's inequality* (also spelled as Tchebysheff's inequality) guarantees that in any probability distribution, "nearly all" values are close to the mean—the precise statement being that no more than $1/k^2$ of the distribution's values can be more than k standard deviations away from the mean.

Let $X$ be a random variable with finite expected value $\mu$ and finite nonzero variance $\sigma^2$. Then for any real number $k > 0$,

$$P\{|X - \mu| \geq k\sigma\} \leq \frac{1}{k^2}.$$

**Definition 5.2.14 (Markov's inequality).** In probability theory, *Markov's inequality* gives an upper bound for the probability that a nonnegative function of a random variable is greater than or equal to some positive constant. It is named after the Russian mathematician Andrey Markov.

If $X$ is any nonnegative random variable and any $a$ in $(0, \infty)$, then

$$P\{X \geq a\} \leq \frac{1}{a}EX.$$

**Definition 5.2.15 (Infinitely often (I.O.)).** Let $(A_n)_{n \geq 1}$ be a sequence of events. Then $\lim_{n \to \infty} A_n = \{w : w \in A_n$ for infinitely many $n\}$, or $\lim_{n \to \infty} A_n = \{w : w \in A_n, \text{I.O.}\}$. Moreover, $\lim_{n \to \infty} A_n = \bigcap_{n=1}^{\infty} \bigcap_{k=n}^{\infty} A_k$.

**Lemma 5.2.16 (Borel-Cantelli Lemma).** *If $(A_n)_{n \geq 1}$ is a sequence of events for which $\sum_{n=1}^{\infty} P\{A_n\} < \infty$, then $P\{A_n, I.O.\} = 0$.*

## 5.3 A-Summability of a Sequence of Random Variables

Let $F$ be the common distribution function of $X_k$s and $X$, a random variable having this distribution. It is also convenient to adopt the convention that $a_{nk} = 0, |a_{nk}|^{-1} = +\infty$. In the next theorem, we study the convergence properties of the sequence

$$Y_n = \sum_{k=0}^{\infty} a_{nk} X_k, \ \text{as} \ n \to \infty.$$

**Theorem 5.3.1.** *A necessary and sufficient condition that* $Y_n \to \mu$ *in probability is that* $\max_{k \in \mathbb{N}} |a_{nk}| \to 0$, *as* $n \to \infty$.

*Proof.* The proof of the sufficiency is very similar to the corresponding argument in [48], but it will be given here for the sake of completeness. First, we have that

$$\lim_{T \to \infty} TP[|X| \geq T] = 0 \tag{5.3.1}$$

since $E|X| < \infty$. Let $X_{nk}$ be $a_{nk}X_k$ truncated at one and $Z_n = \sum_{k=0}^{\infty} X_{nk}$. Now for all $n$ sufficiently large, since $\max_{k \in \mathbb{N}} |a_{nk}| \to 0$, it follows from (5.3.1) that

$$P[Z_n \neq Y_n] \leq \sum_{k=0}^{\infty} P[X_{nk} \neq a_{nk}X_k] = \sum_{k=0}^{\infty} P[|X| \geq \frac{1}{|a_{nk}|}] \leq \epsilon \sum_{k=0}^{\infty} |a_{nk}| \leq \epsilon M.$$

It will therefore suffice to show that $Z_n \to \mu$ in probability. Note that

$$\lim_{n \to \infty} [EZ_n - \mu] = \lim_{n \to \infty} \left[ \sum_{k=0}^{\infty} a_{nk} \left( \int_{|x| < |a_{nk}|^{-1}} x dF - \mu \right) + \mu \left( \sum_{k=0}^{\infty} a_{nk} - 1 \right) \right] = 0.$$

Since

$$\frac{1}{T} \int_{|x| < T} x^2 dF = \frac{1}{T} \left\{ -T^2 P[|x| \geq T] + 2 \int_0^T x P[|x| \geq x] dx \right\} \to 0,$$

it follows that for all $n$ sufficiently large

$$\sum_{k=0}^{\infty} \text{Var } X_{nk} \leq \sum |a_{nk}|^2 \int_{|x| < |a_{nk}|^{-1}} x^2 dF \leq \epsilon \sum_{k=0}^{\infty} |a_{nk}| \leq \epsilon M. \tag{5.3.2}$$

But $E(\sum_{k=0}^{\infty} |X_{nk}|)^2$ is easily seen to be finite so that $\text{Var } Z_n = \sum_{k=0}^{\infty} \text{Var } X_{nk}$ which tends to zero by (5.3.2). An application of Chebyshev's inequality completes the proof of sufficiency. For necessity, let $U_k = X_k - \mu$, $T_n = \sum_{k=0}^{\infty} a_{nk} U_k$ so that $T_n \to 0$ in probability and hence in law. Let $g(u) = E e^{iuU_k}$ be the characteristic function of $U_k$. We have that $\prod_{k=1}^{\infty} g(a_{nk}u) \to 1$ as $n \to \infty$. But

$$\left| \prod_{k=1}^{\infty} g(a_{nk}u) \right| \leq |g(a_{nm}u)| \leq 1$$

for any $m$, so that for any sequence $k_n$,

$$\lim_{n \to \infty} |g(a_{n,k_n}u)| = 1. \tag{5.3.3}$$

Since $U_k$ is nondegenerate, there is a $u_0$ such that $|g(u)| < 1$ for $0 < |u| < u_0$ [57, p. 202]. Letting $u = u_0/2M$, it follows that $|a_{n,k_n} u| \le Mu = u_0/2$ and then $a_{n,k_n} u \to 0$, as $n \to \infty$, by (5.3.3). Choosing $k_n$ to satisfy $|a_{n,k_n}| = \max_{k \in \mathbb{N}} |a_{nk}|$.

This completes the proof of Theorem 5.3.1.                                    □

In Theorem 5.3.1 excluding the trivial case when $X_k$ is almost surely equal to $\mu$, it has been shown that $Y_n \to \mu$ in probability if and only if $\max_{k \in \mathbb{N}} |a_{nk}| \to 0$. This condition is not enough, however, to guarantee almost sure (a.s.) convergence. To obtain this the main result is proved in the following theorem [56].

**Theorem 5.3.2.** *If* $\max_{k \in \mathbb{N}} |a_{nk}| = O(n^{-\gamma})$, $\gamma > 0$, *then* $E|X_k|^{1+\frac{1}{\gamma}} < \infty$ *implies that* $Y_n \to \mu$ *a.s.*

For the proof of Theorem 5.3.2, we need the following lemmas.

**Lemma 5.3.3 ([81, Lemma 1]).** *If* $E|X|^{1+\frac{1}{\gamma}} < \infty$ *and* $\max_{k \in \mathbb{N}} |a_{nk}| \le Bn^{-\gamma}$, *then for every* $\epsilon > 0$,

$$\sum_{n=0}^{\infty} P[|a_{nk} X_k| \ge \epsilon, \text{ for some } k] < \infty$$

*Proof.* It suffices to consider $B = 1$ and $\epsilon = 1$ for both the matrix $A$ and the random variables $X_k$ may be multiplied by a positive constant if necessary. (Assumption (1.3.2) is not used in this proof). Let

$$N_n(x) = \sum_{[k:|a_{nk}|^{-1} \le x]} |a_{nk}|.$$

Notice that $N_n(x) = 0$, for $x < n^\gamma$, and $\int_0^\infty dN_n(x) = \sum_{k=0}^{\infty} |a_{nk}| \le M$. If $G(x) = P[|x| \ge x]$, $\lim TG(t) = 0$, as $T \to \infty$ since $E|X| < \infty$, and thus

$$\sum_{k=0}^{\infty} P[|a_{nk} X_k| \ge 1] = \sum_{k=0}^{\infty} G(|a_{nk}|^{-1})$$

$$= \int_0^\infty XG(x) dN_n(x)$$

$$= \lim_{T \to \infty} TG(T) N_n(T) - \int_0^\infty N_n(\bar{x}) d[xG(x)]$$

$$\le M \int_{n^\gamma}^\infty d|XG(x)|. \tag{5.3.4}$$

To estimate the last integral, observe that, for $z < y$,

$$yG(y) - zG(z) = (y - z)G(z) + y[G(y) - G(z)],$$

so that

$$\int_{n^\gamma}^\infty d|xG(x)| = \sum_{j=n}^\infty \int_{j^\gamma}^{(j+1)^\gamma} d|xG(x)|$$

$$\leq \sum_{j=n}^\infty [(j+1)^\gamma - j^\gamma]G(j^\gamma) + \sum_{j=n}^\infty (j+1)^\gamma [G(j^\gamma) - G((j+1)^\gamma)].$$

Summing the first of the final series by parts and using the existence of $E|X|$, we see that it is dominated by the second series, and thus

$$\int_{n^\gamma}^\infty d|xG(x)| \leq 2\sum_{j=n}^\infty (j+1)^\gamma [G(j^\gamma) - G((j+1)^\gamma)]. \tag{5.3.5}$$

Finally, by (5.3.4) and (5.3.5),

$$\sum_{n=1}^\infty P[|a_{nk}X_k| \geq 1 \text{ for } k] \leq \sum_{n=1}^\infty \sum_{k=1}^\infty P[|a_{nk}X_k| \geq 1]$$

$$\leq 2M \sum_{n=1}^\infty \sum_{j=n}^\infty (j+1)^\gamma [G(j^\gamma) - G((j+1)^\gamma)]$$

$$= 2M \sum_{j=1}^\infty j(j+1)^\gamma [G(j^\gamma) - G((j+1)^\gamma)]$$

$$\leq 2^{\gamma+1}M \int |x|^{1+\frac{1}{\gamma}} dF(x) < \infty.$$

This completes the proof of Lemma 5.3.3.                                        □

**Lemma 5.3.4 ([81, Lemma 2]).** *If $E|X|^{1+\frac{1}{\gamma}} < \infty$ and $\max_{k\in\mathbb{N}} |a_{nk}| \leq Bn^{-\gamma}$, then, for $\alpha < \gamma/2(\gamma+1)$,*

$$\sum_{n=0}^\infty P[|a_{nk}X_k| \geq n^{-\alpha}, \text{ for at least two values of } k] < \infty.$$

*Proof.* By the Markov's inequality,

$$\sum_{n=0}^\infty P[|a_{nk}X_k| \geq n^{-\alpha}] \leq |a_{nk}|^{1+\frac{1}{\gamma}} E|X|^{1+\frac{1}{\gamma}} n^{\alpha(1+\frac{1}{\gamma})},$$

so that

$$P[|a_{nk}X_k| \geq n^{-\alpha} \text{ for at least two } k]$$

$$\leq \sum_{j \neq k} P[|a_{nj}X_j| \geq n^{-\alpha}, |a_{nk}X_k| \geq n^{-\alpha}]$$

$$\leq (E|X|^{1+\frac{1}{\gamma}})^2 n^{2\alpha(1+\frac{1}{\gamma})} \sum_{j \neq k} |a_{nj}|^{1+\frac{1}{\gamma}} |a_{nk}|^{1+\frac{1}{\gamma}}$$

$$\leq (E|X|^{1+\frac{1}{\gamma}})^2 B^{2/\gamma} M^2 n^{2[-1+\alpha(1+\frac{1}{\gamma})]},$$

and the final estimate will converge when summed on $n$ provided that $\alpha < \gamma/2(\gamma+1)$.

This completes the proof of Lemma 5.3.4. $\qquad\qquad\square$

**Lemma 5.3.5 ([81, Lemma 3]).** *If* $\mu = 0$, $E|X|^{1+\frac{1}{\gamma}} < \infty$, *and* $\max_{k \in \mathbb{N}} |a_{nk}| \leq Bn^{-\gamma}$, *then for every* $\epsilon > 0$,

$$\sum_{n=0}^{\infty} P\left[\left|{\sum_{k}}' a_{nk}X_k\right| \geq \epsilon\right] < \infty,$$

*where*

$${\sum_{k}}' a_{nk}X_k = \sum_{[k:|a_{nk}X_k|<n^{-\alpha}]} a_{nk}X_k,$$

*and* $0 < \alpha < \gamma$.

*Proof.* Let $X_{nk} = \begin{cases} X_k, & |a_{nk}X_k| < n^{-\alpha}, \\ 0, & \text{otherwise} \end{cases}$ and $\beta_{nk} = EX_{nk}$. If $a_{nk} = 0$ then $\beta_{nk} = \mu = 0$, while if $a_{nk} \neq 0$, then

$$|\beta_{nk}| = \left|\mu - \int_{|x| \geq n^{-\alpha}|a_{nk}|^{-1}} x \, dF\right| \leq \int_{|x| \geq n^{-\alpha}B^{-1}n^{\gamma}} |x| \, dF.$$

Therefore $\beta_{nk} \to 0$, uniformly in $k$ and $\sum_{k=0}^{\infty} a_{nk}\beta_{nk} \to 0$.

Let $Z_{nk} = X_{nk} - \beta_{nk}$, so that $E|Z_{nk}| = 0$; $E|Z_{nk}|^{1+\frac{1}{\gamma}} \leq C$, for some $C$, and $|a_{nk}Z_{nk}| \leq 2n^{-\alpha}$. Now

$${\sum_{k}}' a_{nk}X_k = \sum_{k=0}^{\infty} a_{nk}X_{nk} = \sum_{k=0}^{\infty} a_{nk}Z_{nk} + \sum_{k} a_{nk}\beta_{nk}.$$

and so for $n$ sufficiently large,

$$\left(\left|\sum_k{}' a_{nk} X_k\right| \geq \epsilon\right) \subseteq \left(\left|\sum_{k=0}^{\infty} a_{nk} Z_{nk}\right| \geq \frac{\epsilon}{2}\right).$$

It will suffice, therefore, to show that

$$\sum_{n=0}^{\infty} P\left(\left|\sum_{k=0}^{\infty} a_{nk} Z_{nk}\right| \geq \epsilon\right) < \infty. \tag{5.3.6}$$

Let $\nu$ be the least integer greater than $1/\gamma$. The necessary estimate will be obtained by computing $E(\sum_{k=0}^{\infty} |a_{nk} Z_{nk}|)^{2\nu}$ which is finite so that

$$E\left(\sum_{k=0}^{\infty} |a_{nk} Z_{nk}|\right)^{2\nu} = \sum_{k_1 \cdots k_{2\nu}} E \prod_{j=1}^{2\nu} a_{n,k_j} Z_{n,k_j}.$$

There is no contribution to the sum on the right so long as there is a $j$ with $k_j \neq k_i$, for all $i \neq j$, since the $Z_{nk}$ are independent and $E Z_{nk} = 0$. The general term to be considered then will have

$$q_1 \text{ of the } k's = \xi_1, \ldots, q_m \text{ of the } k's = \xi_m,$$

$$r_1 \text{ of the } k's = \eta_1, \ldots, r_p \text{ of the } k's = \eta_p,$$

where $2 \leq q_i \leq 1 + \frac{1}{\gamma}$, $r_j > 1 + \frac{1}{\gamma}$, and

$$\sum_{i=1}^{m} q_i + \sum_{j=1}^{p} r_j = 2\nu.$$

Then,

$$E \prod_{i=1}^{m} (a_{n,\xi_i} Z_{n\xi_i})^{q_i} \prod_{j=1}^{p} (a_{n,\eta_j} Z_{n,\eta_j})^{r_j}$$

$$\leq (1+c)^{\nu} \prod_{i=1}^{m} |a_{n,\xi_i}|^{q_i} \prod_{j=1}^{p} |a_{n,\eta_j}|^{1+\frac{1}{\gamma}} (2n^{-\alpha})^{\left(r_j-1-\frac{1}{\gamma}\right)}$$

$$\leq (1+c)^{\nu} \prod_{i=1}^{m} |a_{n,\xi_i}| \prod_{j=1}^{p} |a_{n,\eta_j}| (Bn^{-\gamma})^{\sum_{i=1}^{m}(q_i-1+\frac{p}{\gamma})} \left(\frac{2}{n^{\alpha}}\right)^{\sum_{j=1}^{\nu}(r_j-1-\frac{1}{\gamma})}, \tag{5.3.7}$$

where $c$ is the upper bound for $E|Z_{nk}|^{1+\frac{1}{\gamma}}$ mentioned above. Now, the power to which $n$ is raised is the negative of

$$\gamma \sum_{i=1}^{m} (q_i - 1) + p + \alpha \sum_{j=1}^{p} \left(r_j - 1 - \frac{1}{\gamma}\right).$$

Now, if $p$ is one (or larger),

$$p + \alpha \sum_{j=1}^{p} \left( r_j - 1 - \frac{1}{\gamma} \right) \geq 1 + \alpha \left( \nu - \frac{1}{\gamma} \right),$$

while if $p = 0$,

$$\gamma \sum_{i=1}^{m} (q_i - 1) = \gamma(2\nu - m) \geq \gamma^{\nu} = 1 + \gamma \left( \nu - \frac{1}{\gamma} \right) \geq 1 + \alpha \left( \nu - \frac{1}{\gamma} \right);$$

the first inequality being a result of

$$m \leq \frac{1}{2} \sum_{i=1}^{m} q_i = \nu.$$

Therefore the expectation in (5.3.7) is bounded by

$$k_1 \prod_{i=1}^{m} |a_{n,\xi_i}| \prod_{j=1}^{p} |a_{n,\eta_j}| n^{-1-\alpha(\nu-\frac{1}{\gamma})}$$

and $k_1$ depends only on $c$, $\gamma$, and $B$. It follows that

$$E \left( \sum_{k=0}^{\infty} a_{nk} Z_{nk} \right)^{2\nu} \leq k_2 n^{-1-\alpha(\nu-\frac{1}{\gamma})}$$

for some $k_2$ which may depend on $c, \gamma, B$, and $M$ but is independent of $n$. An application of the Markov's inequality now yields (5.3.6).

This completes the proof of Lemma 5.3.5. □

*Proof of Theorem 5.3.2.* Observe that

$$\sum_{k=0}^{\infty} a_{nk} X_k = \sum_{k=0}^{\infty} a_{nk}(X_k - \mu) + \mu \sum_{k=0}^{\infty} a_{nk}$$

and the last term converges to $\mu$ by (1.3.3). Therefore, we may consider only the case $\mu = 0$. By the Borel-Cantelli Lemma, it suffices to show that for every $\epsilon > 0$,

$$\sum_{n=0}^{\infty} P \left( \left| \sum_{k=0}^{\infty} a_{nk} X_k \right| \geq \epsilon \right) < \infty. \tag{5.3.8}$$

But

$$\left(\left|\sum_{k=0}^{\infty} a_{nk} X_k\right| \geq \epsilon\right) \subset \left(\left|\sum_{k=0}^{\infty} a_{nk} X_k\right| \geq \frac{\epsilon}{2}\right)$$

$$\cup \left(|a_{nk} X_k| \geq \frac{\epsilon}{2} \text{ for some } k\right)$$

$$\cup \left(|a_{nk} X_k| \geq n^{-\alpha} \text{ for at least two } k\right).$$

Now if $0 < \alpha < \gamma/2(\gamma + 1)$, then $\alpha < \gamma$ also and the series (5.3.8) converges as a consequence of Lemma 5.3.3–5.3.5.

This completes the proof of Theorem 5.3.2.                                    □

## 5.4  Strong Law of Large Numbers

In the next theorem, we study the problems arising out of the *strong law of large numbers*.

In probability theory, the *law of large numbers* (LLN) is a theorem that describes the result of performing the same experiment in a large number of times. According to the law, the average of the results obtained from a large number of trials should be close to the expected value and will tend to become closer as more trials are performed.

The *strong law of large numbers* states that the sample average converges almost surely to the expected value ($X_n \to \mu(C, 1)$ a.s., as $n \to \infty$), i.e.,

$$P\left[\lim_{n\to\infty} \frac{X_1 + X_2 + \cdots + X_n}{n} = \mu\right] = 1.$$

Kolmogorov's strong law of large numbers asserts that $EX_1 \to \mu$ if and only if $\sum W_i$ is a.e. $(C, 1)$-summable to $\mu$, i.e., the $(C, 1)$-limit of $(X_n)$ is $\mu$ a.e. By the well-known inclusion theorems involving Cesàro and Abel summability (cf. [41], Theorems 43 and 55), this implies that $\sum W_i$ is a.e. $(C, \alpha)$-summable to $\mu$ for any $\alpha \geq 1$ and that $\sum W_i$ is a.e. $(A)$-summable to $\mu$; where $W_n = X_n - X_{n-1}$ ($X_0 = W_0 = 0$). In fact, the converse also holds in the present case and we have the following theorem.

**Theorem 5.4.1.** *If $X_1, X_2, X_3, \ldots$ is a sequence of i.i.d. random variables and $\alpha \geq 1$ and are given real numbers, then the following statements are equivalent:*

$$E\, X_1 = \mu \tag{5.4.1}$$

$$\lim_{n\to\infty} \frac{X_1 + X_2 + \cdots + X_n}{n} = \mu \ a.e. \tag{5.4.2}$$

$$\lim_{n\to\infty} \sum_{i=1}^{n-1} \frac{\binom{i+\alpha-1}{i} X_{n-1}}{\binom{n+\alpha}{n}} = \mu \ \ a.e.,$$ (5.4.3)

$$where \ \binom{j+\beta}{j} = \frac{(\beta+1)\cdots(\beta+j)}{j!}$$

$$\lim_{\lambda\to 1-} (1-\lambda) \sum_{i=1}^{\infty} \lambda^i X_i = \mu \ \ a.e.$$ (5.4.4)

*Proof.* The implications (5.4.2) $\Rightarrow$ (5.4.3) $\Rightarrow$ (5.4.4) are well known (cf. [41]). We now prove that (5.4.4) implies (5.4.1). By (5.4.4)

$$\lim_{m\to\infty} \frac{1}{m} \sum_{n=1}^{\infty} e^{-n/m} X_n^s = 0 \ (a.e.),$$

where $X_n^s = X_n - X_n'$ with $X_n'$, $n \geq 1$, and $X_n$, $n \geq 1$, being i.i.d. Let

$$Y_m = \frac{1}{m} \sum_{n=1}^{m} e^{-n/m} X_n^s, \ \ Z_m = \frac{1}{m} \sum_{n=m+1}^{\infty} e^{-n/m} X_n^s.$$

Then $Y_m + Z_m \overset{P}{\to} 0$, as $m \to \infty$, $Y_m$ and $Z_m$ are independent and symmetric. Therefore it follows easily from the Levy's inequality [57, p. 247] that $Z_m \overset{P}{\to} 0$. Since $Z_m$ and $(Y_1, \ldots, Y_m)$ are independent and $Y_m + Z_m \to 0$ a.e., $Z_m \overset{P}{\to} 0$, we obtain by Lemma 3 of [23] that $Y_m \to 0$ a.e. Letting $Y_m^{(1)} = Y_m - e^{(m-1}X_m^s)$, since $e^{(m-1}X_m^s) \overset{P}{\to} 0$, we have by Lemma 3 of [10] that $X_m^s/m \to \infty$ a.e. By the Borel-Cantelli lemma, this implies that $E|X_1| < 1$. As established before, we then have $X_n \to EX_1(A)$ and so by (5.4.4), $\mu = EX_1$.

This completes the proof of Theorem 5.4.1. $\qquad \square$

*Remark 5.4.2.* Chow [22] has shown that unlike the Cesàro and Abel methods which require $E|X_1| < \infty$ for summability, the Euler and Borel methods require $EX_1^2 < \infty$ for summability. Specifically, if $X_1, X_2, \ldots$ are i.i.d., then the following statements are equivalent:

$$EX_1 = \mu, \ EX_1^2 < \infty,$$

$X_n \to \mu(E, q)$, for some or equivalently for every $q > 0$, i.e.,

$$\lim_{n\to\infty} \frac{1}{(q+1)^n} \sum_{k=1}^{n} \binom{n}{k} q^{n-k} X_k = \mu \ \ a.e.,$$

$$\lim_{n\to\infty} X_n = \mu(B), \ i.e. \ \lim_{\lambda\to\infty} \frac{1}{e^\lambda} \sum_{k=1}^{\infty} \frac{\lambda^k}{k!} X_k = \mu \ \ a.e..$$

# Chapter 6
# Almost Summability

## 6.1 Introduction

In the theory of sequence spaces, an application of the well-known Hahn-Banach
Extension Theorem gives rise to the notion of Banach limit which further leads to
an important concept of almost convergence. That is, the lim functional defined on
$c$ can be extended to the whole of $\ell_\infty$ and this extended functional is known as the
Banach limit [11]. In 1948, Lorentz [58] used this notion of weak limit to define
a new type of convergence, known as the almost convergence. Since then a huge
amount of literature has appeared concerning various generalizations, extensions,
and applications of this method.

In this chapter, we study the almost conservative and almost regular matrices and
their applications.

## 6.2 Definitions and Notations

First we define almost convergence which will be used to define almost conservative
and almost regular matrices.

**Definition 6.2.1.** A linear functional $L$ on $\ell_\infty$ is said to be a *Banach limit* if it has
the following properties:

(i) $L(x) \geq 0$ if $x \geq 0$;
(ii) $L(e) = 1, e = (1, 1, 1, \ldots)$;
(ii) $L(Sx) = L(x)$; where $S$ is the shift operator defined by $(Sx)_n = x_{n+1}$.

**Definition 6.2.2.** A bounded sequence $x = (x_k)$ is said to be *almost convergent* to
the value $l$ if all its Banach limits coincide, i.e., $L(x) = l$ for all Banach limits $L$.

**Definition 6.2.3.** A sequence $x = (x_k)$ is said to be *almost A-summable* to the
value $l$ if its $A$-transform is almost convergent to $l$.

M. Mursaleen, *Applied Summability Methods*, SpringerBriefs in Mathematics,
DOI 10.1007/978-3-319-04609-9_6, © M. Mursaleen 2014

Lorentz [58] established the following characterization:

A sequence $x = (x_k)$ is *almost convergent* to the number $l$ if and only if $t_{mn}(x) \to l$ as $m \to \infty$ uniformly in $n$, where

$$t_{mn} = t_{mn}(x) = \frac{1}{m+1} \sum_{i=0}^{m} x_{n+i}.$$

The number $l$ is called the *generalized limit of $x$*, and we write $l = F\text{-}\lim x$. We denote the set of all almost convergent sequences by $F$, i.e.,

$$F := \left\{ x \in \ell_\infty : \lim_{m \to \infty} t_{mn}(x) = l \text{ uniformly in } n \right\}.$$

The sequences which are almost convergent are said to be summable by the method $F$, i.e., $x \in F$ we mean $x$ is almost convergent and $F - \lim x = L(x)$.

**Definition 6.2.4.** Let $A = (a_{mk})_{m;k=0}^{\infty}$ be a regular matrix method. A bounded sequence $x = (x_k)$ is said to be $F_A$- *summable* to the value $l$ if $y_{mn} = \sum_{k=0}^{\infty} a_{mk} x_{k+n} \to l$ as $m \to \infty$ uniformly in $n$.

Note that if $A$ is replaced by the $(C, 1)$ matrix, then $F_A$-summability is reduced to almost convergence.

*Example 6.2.5.* The following statements and concepts may be useful:

(i) For $z \in \mathbb{C}$ on the circumference of $|z| = 1$, $L(z^n) = 0$ holds everywhere except for $z = 1$. For from

$$\left| \frac{1}{k} \left( z^n + z^{n+1} + \cdots + z^{n+k-1} \right) \right| = \left| z^n \frac{1 - z^k}{k(1 - z)} \right| \le \frac{2}{k(1 - |z|)},$$

the assertion follows immediately.

It is easy to see [58] that the geometric series $\sum z^n$ for $|z| = 1, z \ne 1$ is almost convergent to $1/(1 - z)$. Hence it follows that the Taylor series of a function $f(z)$, which for $|z| < 1$ is regular and on $|z| = 1$ has simple poles, is almost convergent at every point of the circumference $|z| = 1$ with the limit $f(z)$.

(ii) A periodic sequence $(x_n)$ for which numbers $N$ and $p$ (the period) exist such that $x_{n+p} = x_n$ holds for $n \ge N$ is almost convergent to the value $L(x_n) = \frac{1}{p}(x_N + x_{N+1} + \cdots + x_{N+p-1})$. For example, the periodic sequence $(1, 0, 0, 1, 0, 0, 1, \ldots)$ is almost convergent to $1/3$.

(iii) We say that a sequence $(x_n)$ is *almost periodic* if for every $\varepsilon > 0$, there are two natural numbers $N$ and $r$, such that in every interval $(k, k + r), k > 0$, at least one "$\varepsilon$-period" $p$ exists. More precisely $|x_{n+p} - x_n| < \varepsilon$, for $n \ge N$ must hold for this $p$. Thus it is easy [58] to see that every almost periodic sequence is almost convergent. But there are almost convergent sequences which are not almost periodic. For example, the sequence $x = (x_k)$ defined by

$$x_k = \begin{cases} 1 \, , k = n^2, \\ 0 \, , k \neq n^2; \, n \in \mathbb{N}. \end{cases}$$

is almost convergent (to 0) but not almost periodic.

*Remark 6.2.6.* The following statements hold [58]:

(1) Note that $c \subset F$ and for $x \in c$, $F - \lim x = \lim x$. That is, every convergent sequence is almost convergent to the same limit but not conversely. For example, the sequence $x = (x_k)$ defined by

$$x_k = \begin{cases} 1 \, , k \text{ is odd}, \\ 0 \, , k \text{ is even}. \end{cases}$$

is not convergent but it is almost convergent to $1/2$.

(2) In contrast to the well-known fact that $c$ is a separable closed subspace of $(\ell_\infty, \| \cdot \|_\infty)$, $F$ is a non-separable closed subspace of $(\ell_\infty, \| \cdot \|_\infty)$.
(3) $F$ is a BK-space with $\| \cdot \|_\infty$.
(4) $F$ is nowhere dense in $\ell_\infty$, dense in itself, and closed and therefore perfect.
(5) The method is not strong in spite of the fact that it contains certain classes of matrix methods for bounded sequences.
(6) Most of the commonly used matrix methods contain the method $F$, e.g., every almost convergent sequence is also $(C, \alpha)$ and $(E, \alpha)$-summable ($\alpha > 0$) to its $F$-limit.
(7) The method $F$ is equivalent to none of the matrix methods, i.e., the method $F$ cannot be expressed in the form of a matrix method.
(8) The method $F$ seems to be related to the Cesàro method $(C, 1)$. In fact the method $(C, 1)$ can be replaced in this definition by any other regular matrix method $A$ satisfying certain conditions.
(9) Since $c \subset F \subset \ell_\infty$, we have $\ell_1 = \ell_\infty^\dagger \subset F^\dagger \subset c^\dagger = \ell_1$. That is, the $\dagger$-dual of $F$ is $\ell_1$, where $\dagger$ stands for $\alpha$, $\beta$, and $\gamma$.

## 6.3  Almost Conservative and Almost Regular Matrices

King [50] used the idea of almost convergence to study the almost conservative and almost regular matrices.

**Definition 6.3.1.** A matrix $A = (a_{nk})_{n;k=1}^\infty$ is said to be *almost conservative* if $Ax \in F$ for all $x \in c$. In addition if $F\text{-}\lim Ax = \lim x$ for all $x \in c$, then $A$ is said to be *almost regular*.

**Theorem 6.3.2.** *The following statements hold:*

(a) $A = (a_{nk})_{n;k=1}^\infty$ *is almost conservative, i.e.,* $A \in (c, F)$ *if and only if (1.3.1) holds and*

$$\exists \alpha_k \in \mathbb{C} \ni \lim_{p \to \infty} t(n, k, p) = \alpha_k \ \text{for each } k \text{ uniformly in } n; \quad (6.3.1)$$

$$\exists \alpha \in \mathbb{C} \ni \lim_{p \to \infty} \sum_{k=0}^{\infty} t(n, k, p) = \alpha \ \text{for each } k \text{ uniformly in } n; \quad (6.3.2)$$

$$\text{where } t(n, k, p) = \frac{1}{p+1} \sum_{j=n}^{n+p} a_{jk} \ \text{for all } n, k, p \in \mathbb{N}.$$

*In this case, the F-limit of Ax is*

$$(\lim x) \left( \alpha - \sum_{k=0}^{\infty} \alpha_k \right) + \sum_{k=0}^{\infty} \alpha_k x_k,$$

*for every* $x = (x_k) \in c$.

(b)  *A is almost regular if and only if the conditions (1.3.1), (6.3.1), and (6.3.2) hold with* $\alpha_k = 0$ *for each k and* $\alpha = 1$, *respectively.*

*Proof.* (a) Let the conditions (1.3.1), (6.3.1), and (6.3.2) hold, and $x = (x_k) \in c$. For every positive integer $n$, set

$$t_{pn}(x) = \frac{1}{p+1} \sum_{k=1}^{\infty} \sum_{j=n}^{n+p} a_{jk} x_k.$$

Then we have

$$|t_{pn}(x)| \leq \frac{1}{p+1} \sum_{k=1}^{\infty} \sum_{j=n}^{n+p} |a_{jk}| |x_k| \leq \|A\| \|x\|_{\infty}.$$

Since $t_{pn}$ is obviously linear on $c$, it follows that $t_{pn} \in c^*$, the continuous dual of $c$, and that $\|t_{pn}\| \leq \|A\|$.

Now

$$t_{pn}(e) = \frac{1}{p+1} \sum_{k=1}^{\infty} \sum_{j=n}^{n+p} a_{jk} = \frac{1}{p+1} \sum_{j=n}^{n+p} \sum_{k=1}^{\infty} a_{jk},$$

so $\lim_p t_{pn}(e)$ exists uniformly in $n$ and equals to $\alpha$. Similarly, $t_{pn}(e^{(k)}) \to \alpha_k$, as $p \to \infty$ for each $k$, uniformly in $n$. Since $\{e, e^{(1)}, e^{(2)}, \ldots\}$ is a fundamental set in $c$, and $\sup_p |t_{pn}(x)| < \infty$ for each $x \in c$, it follows that $\lim_p t_{pn}(x) = t_n(x)$ exists for all $x \in c$. Furthermore, $\|t_n\| \leq \liminf_p \|t_{pn}\|$ for each $n$, and $t_n \in c^*$. Thus,

$$t_n(x) = (\lim x) \left[ t_n(e) - \sum_{k=0}^{\infty} t_n\left(e^{(k)}\right) \right] + \sum_{k=0}^{\infty} x_k t_n\left(e^{(k)}\right)$$

$$= (\lim x) \left( \alpha - \sum_{k=0}^{\infty} \alpha_k \right) + \sum_{k=0}^{\infty} x_k \alpha_k,$$

an expression independent of $n$. Denote this expression by $L(x)$.

In order to see that $\lim_p t_{pn}(x) = L(x)$ uniformly in $n$, set $F_{pn}(x) = t_{pn}(x) - L(x)$. Then $F_{pn} \in c^*$, $\|F_{pn}\| \leq 2\|A\|$ for all $p$ and $n$, $\lim_p F_{pn}(e) = 0$ uniformly in $n$, and $\lim_p F_{pn}(e^{(k)}) = 0$ uniformly in $n$ for each $k$. Let $K$ be an arbitrary positive integer. Then

$$x = (\lim x)e + \sum_{k=1}^{K} (x_k - \lim x)\, e^{(k)} + \sum_{k=K+1}^{\infty} (x_k - \lim x)\, e^{(k)},$$

and we have

$$F_{pn}(x) = (\lim x) F_{pn}(e) + \sum_{k=1}^{K} (x_k - \lim x)\, F_{pn}\left(e^{(k)}\right)$$

$$+ F_{pn} \left[ \sum_{k=K+1}^{\infty} (x_k - \lim x)\, e^{(k)} \right].$$

Now,

$$\left| F_{pn} \left[ \sum_{k=K+1}^{\infty} (x_k - \lim x)\, e^{(k)} \right] \right| \leq 2\|A\| \left( \sup_{k \geq K+1} |x_k - \lim x| \right)$$

for all $p$ and $n$. By first choosing a fixed $K$ large enough, each of the three displayed terms for $F_{pn}(x)$ can be made to be uniformly small in absolute value for all sufficiently large $p$, so $\lim_p F_{pn}(x) = 0$ uniformly in $n$. This shows that $\lim_p t_{pn}(x) = L(x)$ uniformly in $n$. Hence $Ax \in F$ for all $x \in c$ and the matrix $A$ is almost conservative.

Conversely, suppose that $A$ is almost conservative. If $x$ is any null sequence, then $Ax \in F \subset \ell_\infty$, i.e., $A \in (c, \ell_\infty)$. We know that $A \in (c, \ell_\infty)$ if and only if $\|A\| < \infty$. Hence, (1.3.1) follows. Furthermore, since $e^{(k)}$ and $e$ are convergent sequences, $k = 0, 1, \ldots$, $\lim_p t_{pn}(e^{(k)})$ and $\lim_p t_{pn}(e)$ must exist uniformly in $n$. Hence, conditions (6.3.1) and (6.3.2) hold, respectively.

(b) Let $A$ be an almost conservative matrix. For $x \in c$, the $F$-limit of $Ax$ is $L(x)$ which reduces to $\lim x$, since $\alpha = 1$ and $\alpha_k = 0$ for each $k$. Hence, $A$ is an almost regular matrix. Conversely, if $A$ almost regular, then $F - \lim Ae = 1$, $F - \lim Ae^{(k)} = 0$, and $\|A\| < \infty$, as in the proof of Part (a).

This completes the proof of Theorem 6.3.2. $\qquad\square$

*Remark 6.3.3.* Every regular matrix is almost regular (since $c \subset F$) but an almost regular matrix need not be regular. Let $C = (c_{nk})$ be defined by

$$
c_{nk} = \begin{cases} \frac{1+(-1)^n}{n+1} & , 0 \leq k \leq n, \\ 0 & , n < k. \end{cases}
$$

Then the matrix $C$ is almost regular but not regular, since $\lim_{n \to \infty} \sum_{k=0}^{\infty} c_{nk}$ does not exist.

*Remark 6.3.4.* It is known that $E^r$ is regular if and only if $0 < r \leq 1$ [2]. It is natural to ask whether or not there exist values of $r$ for which $E^r$ is almost regular but not regular. But this is not the case for Euler matrix $E^r$. In fact, we have that $E^r$ is almost regular if and only if it is regular [50].

## 6.4   Almost Coercive Matrices

Eizen and Laush [31] considered the class of almost coercive matrices.

**Definition 6.4.1.** A matrix $A = (a_{nk})_{n;k=1}^{\infty}$ is said be *almost coercive* if $Ax \in F$ for all $x \in l_{\infty}$.

**Theorem 6.4.2.** $A = (a_{nk})_{n;k=1}^{\infty}$ *is almost coercive, i.e.,* $A \in (\ell_{\infty}, F)$ *if and only if the conditions (1.3.1) and (6.3.1) hold and*

$$
\exists \alpha_k \in \mathbb{C} \ni \lim_{p \to \infty} \sum_{k=0}^{\infty} |t(n,k,p) - \alpha_k| = 0 \ \text{uniformly in } n. \tag{6.4.1}
$$

*In this case, the $F$-limit of $Ax$ is $\sum_{k=0}^{\infty} \alpha_k x_k$ for every $x = (x_k) \in \ell_{\infty}$.*

*Proof.* Suppose that the matrix $A$ satisfies conditions (1.3.1), (6.3.1), and (6.4.1). For any positive integer $K$,

$$
\sum_{k=1}^{K} |\alpha_k| = \sum_{k=1}^{K} \lim_p \frac{1}{p+1} \left| \sum_{j=n}^{n+p} a_{jk} \right|
$$

$$
= \lim_{p \to \infty} \frac{1}{p+1} \sum_{k=1}^{K} \left| \sum_{j=n}^{n+p} a_{jk} \right|
$$

$$
\leq \limsup_p \frac{1}{p+1} \sum_{j=n}^{n+p} \sum_{k=1}^{\infty} |a_{jk}| \leq \|A\|.
$$

This implies that $\sum_{k=1}^{\infty} |\alpha_k|$ converges and that $\sum_{k=1}^{\infty} \alpha_k x_k$ is defined for every bounded sequence $x \in \ell_\infty$.

Let $x \in \ell_\infty$. For $p \in \mathbb{N}$, define the shift operator $S$ on $\ell_\infty$ by $S^p(x) = x_{n+p}$. Then,

$$\left\| \frac{Ax + S(Ax) + \cdots + S^p(Ax)}{p+1} - \left( \sum_{k=0}^{\infty} \alpha_k x_k \right) e \right\| = \sup_{n \in \mathbb{N}} \left| \sum_{k=1}^{\infty} \sum_{j=n}^{n+p} \frac{a_{jk} - \alpha_k}{p+1} x_k \right|$$

$$\leq \|A\| \left( \sup_{n \in \mathbb{N}} \sum_{k=1}^{\infty} \frac{1}{p+1} \left| \sum_{j=n}^{n+p} a_{jk} - \alpha_k \right| \right).$$

By letting $p \to \infty$ and the uniformity of the limits in condition (6.4.1), it follows that

$$\lim_{p \to \infty} \frac{Ax + S(Ax) + \cdots + S^p(Ax)}{p+1} = \left( \sum_{k=0}^{\infty} \alpha_k x_k \right) e$$

and that $Ax \in F$ with $F - \lim Ax = \sum_{k=0}^{\infty} \alpha_k x_k$. Hence, $A$ is almost coercive.

Conversely, suppose that $A \in (\ell_\infty, F)$. Then $A \in (\ell_\infty, c)$ and so the conditions (1.3.1) and (6.3.1) follow immediately by Theorem 6.3.2. To prove the necessity of (6.4.1), let for some $n$

$$\limsup_p \frac{1}{p+1} \sum_{k=1}^{\infty} \left| \sum_{j=n}^{n+p} (a_{jk} - \alpha_k) \right| = N > 0.$$

Since $\|A\|$ is finite, $N$ is finite also. We observe that since $\sum_{k=1}^{\infty} |\alpha_k| < \infty$, the matrix $B = (b_{nk})$, where $b_{nk} = a_{nk} - \alpha_k$, is also a almost coercive matrix. If one sets $F_{kp} = |\sum_{j=n}^{n+p} (a_{jk} - \alpha_k)|/(p+1)$, and $E_{kt} = F_{k,p_t}$, one can follow the construction in the proof of Theorem 2.1 in [31] to obtain a bounded sequence whose transform by the matrix $B$ is not in $F$. This contradiction shows that the limit in (6.4.1) is zero for every $n$.

To show that this convergence is uniform in $n$, we invoke the following lemma, which is proved in [88]. □

**Lemma 6.4.3.** *Let $\{H(n)\}$ be a countable family of matrices $H(n) = \{h_{pk}(n)\}$ such that $\|H(n)\| \leq M < \infty$ for all $n$ and $h_{pk}(n) \to 0$, as $p \to \infty$ for each $k$, uniformly in $n$. Then, $\sum_{k=0}^{\infty} h_{pk}(n) \to 0$, as $p \to \infty$, uniformly in $n$ for all $x \in \ell_\infty$ if and only if $\sum_{k=0}^{\infty} |h_{pk}(n)| \to 0$, as $p \to \infty$, uniformly in $n$.*

*Proof.* Put $h_{pk}(n) = \sum_{j=n}^{n+p} (a_{jk} - \alpha_k)/(p+1)$ and let $H(n)$ be the matrix $(h_{pk}(n))$. Then $\|H(n)\| \leq 2\|A\|$ for every $n$, and that $\lim_p h_{pk}(n) = 0$ for each $k$, uniformly in $n$ by the condition (6.3.1). For any $x \in \ell_\infty$, $\lim_p \sum_{k=0}^{\infty} h_{pk}(n) x_k = F - \lim Ax -$

$\sum \alpha_k x_k$, and the limit exists uniformly in $n$ since $Ax \in F$. Moreover this limit is zero since

$$\sum_{k=1}^{\infty} |h_{pk}(n)x_k| \le \|x\| \frac{1}{p+1} \sum_{k=0}^{\infty} |\sum_{j=n}^{n+p} [a_{jk} - \alpha_k]|.$$

Thus, $\lim_p \sum_{k=0}^{\infty} |h_{pk}(n)| = 0$ uniformly in $n$, and the matrix $A$ satisfies the condition (6.4.1).

This completes the proof.                                                                     □

*Remark 6.4.4.* The classes of almost regular and almost coercive matrices are disjoint ([89], Theorem 4).

**Definition 6.4.5.** A matrix $A$ is said to be *strongly regular (cf. [58])* if it sums all almost convergent sequences and $\lim Ax = F\text{-}\lim x$ for all $x \in F$.

**Theorem 6.4.6.** *A is strongly regular if and only if A is regular and*

$$\lim_{n \to \infty} \sum_{k=0}^{\infty} |a_{nk} - a_{n,k+1}| = 0.$$

Duran [29] considered the class of almost strongly regular matrices.

**Definition 6.4.7.** A matrix $A = (a_{nk})_{n;k=1}^{\infty}$ is said be *almost strongly regular* if $Ax \in F$ for all $x \in F$.

**Theorem 6.4.8.** *$A = (a_{nk})_{n;k=1}^{\infty}$ is almost strongly regular if and only if A is almost regular, and*

$$\lim_{p \to \infty} \sum_{k=0}^{\infty} |t(n, k, p) - t(n, k+1, p)| = 0 \quad \text{uniformly in } n.$$

# Chapter 7
# Almost Summability of Taylor Series

## 7.1 Introduction

In Chap. 4, we applied the generalized Lototski or $[F, d_n]$-summability to study
the regions in which this method sums a Taylor series to the analytic continuation
of the function which it represents. In the applications of summability theory to
function theory it is important to know the region in which the sequence of partial
sums of the geometric series is $A$-summable to $1/(1-z)$ for a given matrix $A$. The
well-known theorem of Okada [78] gives the domain in which a matrix $A = (a_{jk})$
sums the Taylor series of an analytic function $f$ to one of its analytic continuations,
provided that the domain of summability of the geometric series to $1/(1-z)$ and
the distribution of the singular points of $f$ are known. In this chapter, we replace
the $[F, d_n]$-matrix or the general Toeplitz matrix by almost summability matrix
to determine the set on which the Taylor series is almost summable to $f(z)$ (see
[51]). Most of the basic definitions and notations of this chapter are already given in
Chap. 4; in fact, this chapter is in continuation of Chap. 4.

## 7.2 Geometric Series

The following theorem is helpful in determining the region in which the sequence
of partial sums of the geometric series is almost $A$-summable to $1/(1-z)$.

**Theorem 7.2.1.** *Let $D$ be a set of complex numbers with $0 \in D^0$ and $1 \notin D$. Let*
$s = \{s_k(z)\}_k = (\sum_{n=0}^{k} z^n)_k$ *denote the sequence of partial sums of the geometric*
*series. Then, $s$ is almost $A$-summable to $1/(1-z)$ uniformly on each compact subset*
*of $D$ if and only if*

$$\lim_{p \to \infty} \frac{1}{p+1} \sum_{j=n}^{n+p} \sum_{k=0}^{\infty} a_{jk} = 1, \text{ uniformly in } n, \tag{7.2.1}$$

M. Mursaleen, *Applied Summability Methods*, SpringerBriefs in Mathematics,
DOI 10.1007/978-3-319-04609-9_7, © M. Mursaleen 2014

$$\lim_{p\to\infty} \frac{1}{p+1} \sum_{j=n}^{n+p} \sum_{k=0}^{\infty} a_{jk} z^{k+1} = 0, \; \textit{uniformly in } n, \qquad (7.2.2)$$

*uniformly on each compact subset of D.*

*Proof.* Suppose that (7.2.1) and (7.2.2) hold. Then,

$$t_{pn}(z) = \frac{1}{p+1} \sum_{j=n}^{n+p} \sum_{k=0}^{\infty} a_{jk} s_k(z)$$

$$= \frac{1}{p+1} \sum_{j=n}^{n+p} \sum_{k=0}^{\infty} a_{jk} \frac{1-z^{k+1}}{1-z}.$$

Hence,

$$\lim_{p\to\infty} t_{pn}(z) = \frac{1}{1-z} - \lim_{p\to\infty} \frac{z}{(p+1)(1-z)} \sum_{j=n}^{n+p} \sum_{k=0}^{\infty} a_{jk} z^k.$$

Therefore,

$$\lim_{p\to\infty} t_{pn}(z) = \frac{1}{1-z},$$

uniformly in $n$ and uniformly on each compact subset of $D$. Hence, $s$ is almost $A$-summable to $1/(1-z)$ uniformly on each compact subset of $D$.

Conversely, suppose that $t_{pn}(z) \to 1/(1-z)$, as $p \to \infty$, uniformly in $n$ and uniformly on each compact subset $K$ of $D$. Then, $t_{pn}(0) \to 1$, as $p \to \infty$, uniformly in $n$. Hence,

$$\lim_{p\to\infty} \frac{1}{p+1} \sum_{j=n}^{n+p} \sum_{k=0}^{\infty} a_{jk} = 1, \; \text{uniformly in } n,$$

i.e., (7.2.1) holds. In particular, we have that the series $\sum_{k=0}^{\infty} a_{jk}$ converges for each $j$. But as above

$$t_{pn}(z) = \frac{1}{p+1} \sum_{j=n}^{n+p} \sum_{k=0}^{\infty} a_{jk} \frac{1-z^{k+1}}{1-z},$$

so that the series $\sum_{k=0}^{\infty} a_{jk}(1-z^{k+1})$ converges for each $j$ and on each compact subset $K$ of $D$. Therefore, $\sum_{k=0}^{\infty} a_{jk} z^{k+1}$ converges for each $j$ and on each compact subset $K$ of $D$. This implies that

$$(z - 1)t_{pn}(z) + t_{pn}(0) = \frac{1}{p+1} \sum_{j=n}^{n+p} \sum_{k=0}^{\infty} a_{jk} z^{k+1}$$

and hence,

$$\lim_{p \to \infty} \frac{1}{p+1} \sum_{j=n}^{n+p} \sum_{k=0}^{\infty} a_{jk} z^{k+1} = (z - 1)\frac{1}{1 - z} + 1 = 0,$$

uniformly in $n$ and uniformly on each compact subset $K$ of $D$, i.e., (7.2.2) holds. This completes the proof. □

**Theorem 7.2.2.** *Let $r \neq 0$ be a complex number. The sequence of partial sums of the geometric series is almost $E^r$-summable to $1/(1-z)$ if and only if $z \in \Omega_r$, where*

$$\Omega_r = \left\{ z : \left| z - \frac{r-1}{r} \right| \leq \frac{1}{|r|}, \ z \neq 1 \right\}.$$

*Proof.* It is easy to see that for the Euler matrix $E^r$ (7.2.1) holds, since $\sum_{k=0}^{\infty} e_{jk}^r = 1$. Hence, by Theorem 7.2.1 it is sufficient to show that $\Omega_r = \Omega$, where

$$\Omega = \left\{ z : \lim_{p \to \infty} \frac{1}{p+1} \sum_{j=n}^{n+p} \sum_{k=0}^{\infty} a_{jk} z^k = 0, \text{ uniformly in } n \right\}.$$

In this case

$$\sum_{j=n}^{n+p} \sum_{k=0}^{\infty} e_{jk}^r z^k = \sum_{j=n}^{n+p} \sum_{k=0}^{j} \binom{j}{k} (1 - r)^{j-k} r^k z^k$$

$$= \sum_{j=n}^{n+p} (1 - r + rz)^j$$

$$= (1 - r + rz)^n \frac{1 - (1 - r + rz)^p}{1 - (1 - r + rz)}.$$

Therefore, $z \in \Omega$ if and only if $|1 - r + rz| \leq 1$, $z \neq 1$. Hence, $\Omega_r = \Omega$. In order that the sequence of partial sums be $E^r$-summable to $1/(1 - z)$, it is necessary that $z$ lies strictly inside the disk $\Omega_r$.

This completes the proof. □

## 7.3 Taylor Series

Note that Theorem 4.3.5 is the source of motivation of the following result which is due to King [51] which provides the set on which the Taylor series is almost summable to $f(z)$.

**Theorem 7.3.1.** *Let $P(z)$ be the Taylor series which represents an analytic function $f$ in a neighborhood of the origin. Let $\sigma = \{\sigma_k(z)\}$ be the sequence of partial sums of $P(z)$. Let $\Gamma$ be continuous. Let $M = M(P;\Gamma)$ and let $\Omega = \cap\{wD : w \notin M, w \neq \infty\}$, where $D$ is a $\Gamma$-regular set with $0 \in D^0$. Let $K$ be a compact subset of $\Omega$ such that $d(K(M^c)^{-1}, D^c) = \delta > 0$ and $0 \in K$. If the sequence of partial sums of the geometric series is almost A-summable to $1/(1-z)$ uniformly on each compact subset of $D$, then $\sigma$ is almost A-summable to $P(z;\Gamma)$ uniformly on $K$.*

*Proof.* From Theorem 7.2.1, we get that (7.2.1) and (7.2.2) hold uniformly on each compact subset of $D$. As in the proof of Theorem 4.3.5, it follows that there exists a rectifiable Jordan curve $\gamma$ which satisfies the following conditions:

(i) $\gamma \subset M(P;\Gamma)$;
(ii) $\Gamma\gamma^{-1} \subset D$;
(iii) $K$ lies in the interior of $\gamma$.

By Lemma 4.3.6, it follows that $P(z;\Gamma)$ is holomorphic in $M(P;\Gamma)$.

Let $P(z) = \sum_{k=0}^{\infty} c_k z^k$. Then the conditions (7.2.1) and (7.2.2), the properties of $\gamma$, the fact that $1 \notin D$, and the calculus of residues yield the following relation:

$$
\begin{aligned}
P(z;\Gamma) &= \frac{1}{2\pi i} \int_{\gamma} \frac{P(w;\Gamma)}{w-z} dw \\
&= \frac{1}{2\pi i} \int_{\gamma} \frac{P(w;\Gamma)}{w} \left(\frac{1}{1-\frac{z}{w}}\right) dw \\
&= \frac{1}{2\pi i} \int_{\gamma} \frac{P(w;\Gamma)}{w} \lim_{p\to\infty} \frac{1}{p+1} \sum_{j=n}^{n+p} \sum_{k=0}^{\infty} a_{jk} s_k \left(\frac{z}{w}\right) dw \\
&= \lim_{p\to\infty} \frac{1}{p+1} \sum_{j=n}^{n+p} \sum_{k=0}^{\infty} a_{jk} \frac{1}{2\pi i} \int_{\gamma} \frac{P(w;\Gamma)}{w} s_k \left(\frac{z}{w}\right) dw \\
&= \lim_{p\to\infty} \frac{1}{p+1} \sum_{j=n}^{n+p} \sum_{k=0}^{\infty} a_{jk} \left(c_0 + c_1 z + \cdots + c_k z^k\right) \\
&= \lim_{p\to\infty} \frac{1}{p+1} \sum_{j=n}^{n+p} \sum_{k=0}^{\infty} a_{jk} \sigma_k(z)
\end{aligned}
$$

uniformly in $n$ and uniformly on $K$. Hence, $\sigma$ is almost $A$-summable to $P(z; \Gamma)$ uniformly on $K$.

This completes the proof. $\square$

*Remark 7.3.2.* From Theorems 7.2.2 and 4.3.5, we conclude that the sequence of partial sums of $P(z)$ is almost $E^r$-summable to $P(z; \Gamma)$ uniformly on $K$. An analysis shows that the sets of almost summability are slightly larger than the sets of summability.

# Chapter 8
# Matrix Summability of Fourier and Walsh-Fourier Series

## 8.1  Introduction

In this chapter we apply regular and almost regular matrices to find the sum of derived Fourier series, conjugate Fourier series, and Walsh-Fourier series (see [4] and [69]). Recently, Móricz [67] has studied statistical convergence of sequences and series of complex numbers with applications in Fourier analysis and summability.

## 8.2  Summability of Fourier Series

Let $f$ be $L$-integrable and periodic with period $2\pi$, and let the *Fourier series* of $f$ be

$$\frac{1}{a_0} + \sum_{k=1}^{\infty} (a_k \cos kx + b_k \sin kx). \tag{8.2.1}$$

Then, the *series conjugate* to it is

$$\sum_{k=1}^{\infty} (b_k \cos kx - a_k \sin kx), \tag{8.2.2}$$

and the *derived series* is

$$\sum_{k=1}^{\infty} k (b_k \cos kx - a_k \sin kx). \tag{8.2.3}$$

M. Mursaleen, *Applied Summability Methods*, SpringerBriefs in Mathematics,
DOI 10.1007/978-3-319-04609-9_8, © M. Mursaleen 2014

Let $S_n(x)$, $\tilde{S}_n(x)$, and $S'_n(x)$ denote the partial sums of series (8.2.1), (8.2.2), and (8.2.3), respectively. We write

$$\psi_x(t) = \psi(f, t) = \begin{cases} f(x+t) - f(x-t), & 0 < t \le \pi; \\ g(x), & t = 0 \end{cases}$$

and

$$\beta_x(t) = \frac{\psi_x(t)}{4 \sin \frac{1}{2} t},$$

where $g(x) = f(x+0) - f(x-0)$. These formulae are correct a.e..

**Theorem 8.2.1.** *Let $f(x)$ be a function integrable in the sense of Lebesgue in $[0, 2\pi]$ and periodic with period $2\pi$. Let $A = (a_{nk})$ be a regular matrix of real numbers. Then for every $x \in [-\pi, \pi]$ for which $\beta_x(t) \in BV[0, \pi]$,*

$$\lim_{n \to \infty} \sum_{k=1}^{\infty} a_{nk} S'_k(x) = \beta_x(0+) \tag{8.2.4}$$

*if and only if*

$$\lim_{n \to \infty} \sum_{k=1}^{\infty} a_{nk} \sin\left(k + \frac{1}{2}\right) t = 0 \tag{8.2.5}$$

*for every $t \in [0, \pi]$, where $BV[0, \pi]$ denotes the set of all functions of bounded variations on $[0, \pi]$.*

We shall need the following well-known *Dirichlet-Jordan Criterion for Fourier series* [101].

**Lemma 8.2.2 (Dirichlet-Jordan Criterion for Fourier Series).** *The trigonometric Fourier series of a $2\pi$-periodic function $f$ having bounded variation converges to $[f(x+0) - f(x-0)]/2$ for every $x$ and this convergence is uniform on every closed interval on which $f$ is continuous.*

We shall also need the following result on the weak convergence of sequences in the Banach space of all continuous functions defined on a finite closed interval [11].

**Lemma 8.2.3.** *Let $C[0, \pi]$ be the space of all continuous functions on $[0, \pi]$ equipped with the sup-norm $\|.\|$. Let $g_n \in C[0, \pi]$ and $\int_0^\pi g_n dh_x \to 0$, as $n \to \infty$, for all $h_x \in BV[0, \pi]$ if and only if $\|g_n\| < \infty$ for all $n$ and $g_n \to 0$, as $n \to \infty$.*

*Proof.* We have

$$
S'_k(x) = \frac{1}{\pi} \int_0^\pi \psi_x(t) \left( \sum_{m=1}^{k} m \sin mt \right) dt
$$

$$
= -\frac{1}{\pi} \int_0^\pi \psi_x(t) \frac{d}{dt} \left[ \frac{\sin \left(k + \frac{1}{2}\right) t}{2 \sin \frac{t}{2}} \right] dt
$$

$$
= I_k + \frac{2}{\pi} \int_0^\pi \sin \left(k + \frac{1}{2}\right) t \, d\beta_x(t),
$$

where

$$
I_k = \frac{1}{\pi} \int_0^\pi \beta_x(t) \cos \frac{t}{2} \left[ \frac{\sin \left(k + \frac{1}{2}\right) t}{\sin \frac{t}{2}} \right] dt.
$$

Then,

$$
\sum_{k=1}^{\infty} a_{nk} S'_k(x) = \sum_{k=1}^{\infty} a_{nk} I_k + \frac{2}{\pi} \int_0^\pi L_n(t) \, d\beta_x(t),
$$

where

$$
L_n(t) = \sum_{k=1}^{\infty} a_{nk} \sin \left(k + \frac{1}{2}\right) t.
$$

Since $\beta_x(t)$ is of bounded variation on $[0, \pi]$ and $\beta_x(t) \to \beta_x(0+)$ as $t \to 0$, $\beta_x(t) \cos \frac{t}{2}$ has also the same properties. Hence, by Lemma 8.2.2, $I_k \to \beta_x(0+)$ as $k \to \infty$.

Since the matrix $A = (a_{nk})$ is regular, we have

$$
\lim_{n \to \infty} \sum_{k=1}^{\infty} a_{nk} I_k = \beta_x(0+). \tag{8.2.6}
$$

Now, it is enough to show that (8.2.5) holds if and only if

$$
\lim_{n \to \infty} \int_0^\pi L_n(t) \, d\beta_x(t) = 0. \tag{8.2.7}
$$

Hence, by Lemma 8.2.3, it follows that (8.2.7) holds if and only if

$$
\| L_n(t) \| \leq M \text{ for all } n \text{ and for all } t \in [0, \pi], \tag{8.2.8}
$$

and (8.2.5) holds, where $M$ is a constant. Since (8.2.8) is satisfied by the regularity of $A$, it follows that (8.2.7) holds if and only if (8.2.5) holds. Hence the result follows immediately.

This completes the proof.                                                                    □

Similarly we can prove the following result for almost regularity.

**Theorem 8.2.4.** *Let $f$ be a function integrable in the sense of Lebesgue in $[0, 2\pi]$ and periodic with period $2\pi$. Let $A = (a_{nk})$ be an almost regular matrix of real numbers. Then for every $x \in [-\pi, \pi]$ for which $\beta_x(t) \in BV[0, \pi]$,*

$$\lim_{p \to \infty} \frac{1}{p+1} \sum_{j=n}^{n+p} \sum_{k=1}^{\infty} a_{jk} S_k'(x) = \beta_x(0+) \text{ uniformly in } n$$

*if and only if*

$$\lim_{p \to \infty} \frac{1}{p+1} \sum_{j=n}^{n+p} \sum_{k=1}^{\infty} a_{jk} \sin\left(k + \frac{1}{2}\right) t = 0 \text{ uniformly in } n$$

*for every $t \in [0, \pi]$.*

**Theorem 8.2.5.** *Let $f(x)$ be a function integrable in the sense of Lebesgue in $[0, 2\pi]$ and periodic with period $2\pi$. Let $A = (a_{nk})$ be a regular matrix of real numbers. Then $A$-transform of the sequence $\{k \tilde{S}_k(x)\}$ converges to $g(x)/\pi$, i.e.,*

$$\lim_{n \to \infty} \sum_{k=1}^{\infty} k a_{nk} \tilde{S}_k(x) = \frac{1}{\pi} g(x) \qquad (8.2.9)$$

*if and only if*

$$\lim_{n \to \infty} \sum_{k=0}^{\infty} a_{nk} \cos k t = 0 \qquad (8.2.10)$$

*for every $t \in (0, \pi]$, where each $a_k, b_k \in BV[0, 2\pi]$.*

*Proof.* We have

$$\tilde{S}_n(x) = \frac{1}{\pi} \int_0^{\pi} \psi_x(t) \sin nt \, dt,$$

$$= \frac{g(x)}{n\pi} + \frac{1}{n\pi} \int_0^{\pi} \cos nt \, d\psi_x(t).$$

Therefore

$$\sum_{k=1}^{\infty} k a_{nk} \tilde{S}_k(x) = \frac{g(x)}{\pi} \sum_{k=1}^{\infty} a_{nk} + \frac{1}{\pi} \int_0^{\pi} K_n(t) \, d\psi_x(t), \qquad (8.2.11)$$

where

$$K_n(t) = \sum_{k=1}^{\infty} a_{nk} \cos kt.$$

Now, taking limit as $n \to \infty$ on both sides of (8.2.10) and using Lemma 8.2.3 and regularity conditions of $A$ as in the proof of Theorem 8.2.1, we get the required result. $\qquad \square$

*Remark 8.2.6.* Analogously, we can state and prove Theorem 8.2.4 for almost regular matrix $A$.

## 8.3 Summability of Walsh-Fourier Series

Let us define a sequence of functions $h_0(x), h_1(x), \ldots, h_n(x)$ which satisfy the following conditions:

$$h_0(x) = \begin{cases} 1, & 0 \le x \le \frac{1}{2}, \\ -1, & \frac{1}{2} \le x < 1, \end{cases}$$

$h_0(x + 1) = h_0(x)$ and $h_n(x) = h_0(2^n x)$, $n = 1, 2, \ldots$. The functions $h_n(x)$ are called the *Rademacher's functions*.

The *Walsh functions* are defined by

$$\phi_n(x) = \begin{cases} 1, & n = 0, \\ h_{n_1}(x) h_{n_2}(x) \cdots h_{n_r}(x), & n > 1, \ 0 \le x \le 1 \end{cases}$$

for $n = 2^{n_1} + 2^{n_2} + \cdots + 2^{n_r}$, where the integers $n_i$ are uniquely determined by $n_{i+1} < n_i$.

Let us recall some basic properties of Walsh functions (see [34]). For each fixed $x \in [0, 1)$ and for all $t \in [0, 1)$

(i) $\phi_n(x \dotplus t) = \phi_n(x) \phi_n(t)$,
(ii) $\int_0^1 f(x \dotplus t) dt = \int_0^1 f(t) dt$, and
(iii) $\int_0^1 f(t) \phi_n(x \dotplus t) dt = \int_0^1 f(x \dotplus t) \phi_n(t) dt$,

where $\dotplus$ denotes the operation in the dyadic group, the set of all sequences $s = (s_n)$, $s_n = 0, 1$ for $n = 1, 2, \ldots$ is addition modulo 2 in each coordinate.

Let for $x \in [0, 1)$,

$$J_k(x) = \int_0^x \phi_k(t)dt, \ k = 0, 1, 2, \ldots$$

It is easy to see that $J_k(x) = 0$ for $x = 0, 1$.

Let $f$ be $L$-integrable and periodic with period 1, and let the *Walsh-Fourier series of* $f$ be

$$\sum_{n=1}^{\infty} c_n \phi_n(x),$$

where

$$c_n = \int_0^1 f(x)\phi_n(x)dx$$

are called the *Walsh-Fourier coefficients of* $f$.

The following result is due to Siddiqi [91].

**Theorem 8.3.1.** *Let $A = (a_{nk})$ be a regular matrix of real numbers. Let $z_k(x) = c_k \phi_k(x)$ for an $L$-integrable function $f \in BV[0, 1)$. Then for every $x \in [0, 1)$*

$$\lim_{n \to \infty} \sum_{k=1}^{\infty} a_{nk} z_k(x) = 0$$

*if and only if*

$$\lim_{n \to \infty} \sum_{k=1}^{\infty} a_{nk} J_k(x) = 0,$$

*where $x$ is a point at which $f(x)$ is of bounded variation.*

This can be proved similarly as our next result which is due to Mursaleen [69] in which we use the notion of $F_A$-summability. Recently, Alghamdi and Mursaleen [4] have applied Hankel matrices for this purpose.

**Theorem 8.3.2.** *Let $A = (a_{nk})$ be a regular matrix of real numbers. Let $z_k(x) = c_k \phi_k(x)$ for an $L$-integrable function $f \in BV[0, 1)$. Then for every $x \in [0, 1)$, the sequence $\{z_k(x)\}_k$ is $F_A$-summable to 0 if and only if the sequence $\{J_k(x)\}_k$ is $F_A$-summable to 0, that is,*

$$\lim_{n \to \infty} \sum_{k=1}^{\infty} a_{nk} z_{k+p}(x) = 0, \ \text{uniformly in } p.$$

*if and only if*

$$\lim_{n \to \infty} \sum_{k=1}^{\infty} a_{nk} J_k(x) = 0 \text{ uniformly in } p,$$

*where $x$ is a point at which $f(x)$ is of bounded variation.*

*Proof.* We have

$$z_k(x) = c_k \phi_k(x) = \int_0^1 f(t) \phi_k(t) \phi_k(x) dt,$$

$$= \int_0^1 f(t) \phi_k(x \dotplus t) dt = \int_0^1 f(x \dotplus t) \phi_k(t) dt,$$

where $x \dotplus t$ belongs to the set $\Omega$ of dyadic rationals in $[0, 1)$; in particular each element of $\Omega$ has the form $p/2^n$ for some nonnegative integers $p$ and $n$, $0 \le p < 2^n$. Now, on integration by parts, we obtain

$$z_k(x) = [f(x \dotplus t) J_k(t)]_0^1 - \int_0^1 J_k(t) df(x \dotplus t),$$

$$= - \int_0^1 J_k(t) df(x \dotplus t), \text{ since } J_k(x) = 0 \text{ for } x \in \{0, 1\}.$$

Hence, for a regular matrix $A = (a_{nk})$ and $p \ge 0$, we have

$$\sum_{k=1}^{\infty} a_{nk} z_{k+p}(x) = - \int_0^1 D_{np}(t) \, dh_x(t), \tag{8.3.1}$$

where

$$D_{np}(t) = \sum_{k=1}^{\infty} a_{nk} J_{k+p}(t), \tag{8.3.2}$$

and $h_x(t) = f(x \dotplus t)$. Write, for any $t \in \mathbb{R}$, $g_{np} = (D_{np}(t))$.

Since $A$ is regular (and hence almost regular), it follows that $\|g_{np}\| < \infty$ for all $n$ and $p$, and $g_{np} \to 0$, as $n \to \infty$ pointwise, uniformly in $p$. Hence by Lemma 8.2.3,

$$\int_0^1 D_{np}(t) dh_x(t) \to 0$$

as $n \to \infty$ uniformly in $p$. Now, letting $n \to \infty$ in (8.3.1) and (8.3.2) and using Lemma 8.2.3, we get the desired result.

This completes the proof. $\qquad\qquad\qquad\qquad\qquad\qquad\qquad\qquad\qquad\quad \square$

*Remark 8.3.3.* If we take the matrix $A$ as the Cesàro matrix $(C, 1)$, then we get the following result for almost summability.

**Theorem 8.3.4.** *Let* $A = (a_{nk})$ *be almost regular matrix of real numbers. Let* $z_k(x) = c_k \phi_k(x)$ *for an* $L$ *-integrable function* $f \in BV[0, 1)$. *Then for every* $x \in [0, 1)$

$$F - \lim_{n \to \infty} \sum_{k=1}^{\infty} a_{nk} z_k(x) = 0$$

*if and only if*

$$F - \lim_{n \to \infty} \sum_{k=1}^{\infty} a_{nk} J_k(x) = 0,$$

*where* $x$ *is a point at which* $f$ *is of bounded variation.*

# Chapter 9
# Almost Convergence in Approximation Process

## 9.1 Introduction

Several mathematicians have worked on extending or generalizing the Korovkin's theorems in many ways and to several settings, including function spaces, abstract Banach lattices, Banach algebras, Banach spaces, and so on. This theory is very useful in real analysis, functional analysis, harmonic analysis, measure theory, probability theory, summability theory, and partial differential equations. But the foremost applications are concerned with constructive approximation theory which uses it as a valuable tool. Even today, the development of Korovkin-type approximation theory is far from complete. Note that the first and the second theorems of Korovkin are actually equivalent to the algebraic and the trigonometric version, respectively, of the classical Weierstrass approximation theorem [1]. In this chapter we prove Korovkin type approximation theorems by applying the notion of almost convergence and show that these results are stronger than original ones.

## 9.2 Korovkin Approximation Theorems

Let $F(\mathbb{R})$ denote the linear space of all real-valued functions defined on $\mathbb{R}$. Let $C(\mathbb{R})$ be the space of all functions $f$ continuous on $\mathbb{R}$. We know that $C(\mathbb{R})$ is a normed space with norm

$$\|f\|_\infty := \sup_{x \in \mathbb{R}} |f(x)|, \ f \in C(\mathbb{R}).$$

We denote by $C_{2\pi}(\mathbb{R})$ the space of all $2\pi$-periodic functions $f \in C(\mathbb{R})$ which is a normed space with

$$\|f\|_{2\pi} = \sup_{t \in \mathbb{R}} |f(t)|.$$

M. Mursaleen, *Applied Summability Methods*, SpringerBriefs in Mathematics, DOI 10.1007/978-3-319-04609-9__9, © M. Mursaleen 2014

We write $L_n(f;x)$ for $L_n(f(s);x)$ and we say that $L$ is a positive operator if $L(f;x) \geq 0$ for all $f(x) \geq 0$.

Korovkin type approximation theorems are useful tools to check whether a given sequence $(L_n)_{n \geq 1}$ of positive linear operators on $C[0,1]$ of all continuous functions on the real interval $[0,1]$ is an approximation process. That is, these theorems exhibit a variety of test functions which assure that the approximation property holds on the whole space if it holds for them. Such a property was discovered by Korovkin in 1953 for the functions $1$, $x$, and $x^2$ in the space $C[0,1]$ as well as for the functions $1$, $cos$, and $sin$ in the space of all continuous $2\pi$-periodic functions on the real line.

The classical *Korovkin first and second theorems* state as follows (see [1, 55]):

**Theorem 9.2.1.** *Let $(T_n)$ be a sequence of positive linear operators from $C[0,1]$ into $F[0,1]$. Then $\lim_{n \to \infty} \|T_n(f,x) - f(x)\|_\infty = 0$, for all $f \in C[0,1]$ if and only if $\lim_{n \to \infty} \|T_n(f_i,x) - e_i(x)\|_\infty = 0$, for $i = 0,1,2$, where $e_0(x) = 1$, $e_1(x) = x$, and $e_2(x) = x^2$.*

**Theorem 9.2.2.** *Let $(T_n)$ be a sequence of positive linear operators from $C_{2\pi}(\mathbb{R})$ into $F(\mathbb{R})$. Then $\lim_{n \to \infty} \|T_n(f,x) - f(x)\|_{2\pi} = 0$, for all $f \in C_{2\pi}(\mathbb{R})$ if and only if $\lim_{n \to \infty} \|T_n(f_i,x) - f_i(x)\|_{2\pi} = 0$, for $i = 0,1,2$, where $f_0(x) = 1$, $f_1(x) = \cos x$, and $f_2(x) = \sin x$.*

## 9.3 Korovkin Approximation Theorems for Almost Convergence

The following result is due to Mohiuddine [60]. In [7], such type of result is proved for almost convergence of double sequences.

**Theorem 9.3.1.** *Let $(T_k)$ be a sequence of positive linear operators from $C[a,b]$ into $C[a,b]$ satisfying the following conditions:*

$$F - \lim_{p \to \infty} \|T_k(1,x) - 1\|_\infty = 0, \tag{9.3.1}$$

$$F - \lim_{p \to \infty} \|T_k(t,x) - x\|_\infty = 0, \tag{9.3.2}$$

$$F - \lim_{p \to \infty} \|T_k(t^2,x) - x^2\|_\infty = 0. \tag{9.3.3}$$

*Then for any function $f \in C[a,b]$ bounded on the whole real line, we have*

$$F - \lim_{k \to \infty} \|T_k(f,x) - f(x)\|_\infty = 0.$$

*Proof.* Since $f \in C[a,b]$ and $f$ is bounded on the real line, we have

$$|f(x)| \leq M, \quad -\infty < x < \infty.$$

Therefore,

$$|f(t) - f(x)| \leq 2M, \quad -\infty < t, x < \infty. \tag{9.3.4}$$

Also, we have that $f$ is continuous on $[a, b]$, i.e.,

$$|f(t) - f(x)| < \epsilon, \quad \forall |t - x| < \delta. \tag{9.3.5}$$

Using (9.3.4) and (9.3.5) and putting $\psi(t) = (t - x)^2$, we get

$$|f(t) - f(x)| < \epsilon + \frac{2M}{\delta^2} \psi, \quad \forall |t - x| < \delta.$$

This means

$$-\epsilon - \frac{2M}{\delta^2} \psi < f(t) - f(x) < \epsilon + \frac{2M}{\delta^2} \psi.$$

Now, we operating $T_k(1, x)$ to this inequality since $T_k(f, x)$ is monotone and linear. Hence,

$$T_k(1, x)\left(-\epsilon - \frac{2M}{\delta^2} \psi\right) < T_k(1, x)(f(t) - f(x)) < T_k(1, x)\left(\epsilon + \frac{2M}{\delta^2} \psi\right).$$

Note that $x$ is fixed and so $f(x)$ is a constant number. Therefore,

$$-\epsilon T_k(1, x) - \frac{2M}{\delta^2} T_k(\psi, x) < T_k(f, x) - f(x) T_k(1, x)$$

$$< \epsilon T_k(1, x) + \frac{2M}{\delta^2} T_k(\psi, x). \tag{9.3.6}$$

But

$$T_k(f, x) - f(x) = T_k(f, x) - f(x) T_k(1, x) + f(x) T_k(1, x) - f(x)$$
$$= [T_k(f, x) - f(x) T_k(1, x)] + f(x)[T_k(1, x) - 1]. \tag{9.3.7}$$

Using (9.3.6) and (9.3.7), we have

$$T_k(f, x) - f(x) < \epsilon T_k(1, x) + \frac{2M}{\delta^2} T_k(\psi, x) + f(x)[T_k(1, x) - 1]. \tag{9.3.8}$$

Let us estimate $T_k(\psi, x)$

$$T_k(\psi, x) = T_k\left[(t - x)^2, x\right]$$
$$= T_k(t^2 - 2tx + x^2, x)$$
$$= T_k(t^2, x) + 2x T_k(t, x) + x^2 T_k(1, x)$$
$$= [T_k(t^2, x) - x^2] - 2x[T_k(t, x) - x] + x^2[T_k(1, x) - 1].$$

Using (9.3.8), we obtain

$$T_k(f,x) - f(x) < \epsilon T_k(1,x) + \frac{2M}{\delta^2}\{[T_k(t^2,x) - x^2] + 2x[T_k(t,x) - x]$$

$$+ x^2[T_k(1,x) - 1]\} + f(x)[T_k(1,x) - 1]$$

$$= \epsilon[T_k(1,x) - 1] + \epsilon + \frac{2M}{\delta^2}\{[T_k(t^2,x) - x^2] + 2x[T_k(t,x) - x]$$

$$+ x^2[T_k(1,x) - 1]\} + f(x)[T_k(1,x) - 1].$$

Since $\epsilon$ is arbitrary, we can write

$$T_k(f,x) - f(x) \le \epsilon[T_k(1,x) - 1] + \frac{2M}{\delta^2}\{[T_k(t^2,x) - x^2] + 2x[T_k(t,x) - x]$$

$$+ x^2[T_k(1,x) - 1]\} + f(x)[T_k(1,x) - 1].$$

Now replacing $T_k(\cdot, x)$ by $D_{n,p}(f,x) = \frac{1}{p+1}\sum_{k=n}^{n+p} T_k(\cdot, x)$, we get

$$D_{n,p}(f,x) - f(x) \le \epsilon[D_{n,p}(1,x) - 1] + \frac{2M}{\delta^2}\{[D_{n,p}(t^2,x) - x^2]$$

$$+ 2x[D_{n,p}(t,x) - x] + x^2[D_{n,p}(1,x) - 1]\}$$

$$+ f(x)[D_{n,p}(1,x) - 1],$$

and therefore

$$\|D_{n,p}(f,x) - f(x)\|_\infty \le \left(\epsilon + \frac{2Mb^2}{\delta^2} + M\right)\|D_{n,p}(1,x) - 1\|_\infty$$

$$+ \frac{4Mb}{\delta^2}\|D_{n,p}(t,x) - x\|_\infty + \frac{2M}{\delta^2}\|D_{n,p}(t^2,x) - x^2\|_\infty.$$

Letting $p \to \infty$ and using (9.3.1)–(9.3.3), we get

$$\lim_{p\to\infty}\|D_{n,p}(f,x) - f(x)\|_\infty = 0 \text{ uniformly in } n.$$

This completes the proof of the theorem.                                         □

In the following example we construct a sequence of positive linear operators satisfying the conditions of Theorem 9.3.1, but it does not satisfy the conditions of Theorem 9.2.1.

*Example 9.3.2.* Consider the sequence of classical Bernstein polynomials

$$B_n(f,x) := \sum_{k=0}^{n} f\left(\frac{k}{n}\right)\binom{n}{k}x^k(1-x)^{n-k}; \ 0 \le x \le 1.$$

Let the sequence $(P_n)$ be defined by $P_n : C[0, 1] \to C[0, 1]$ with $P_n(f, x) = (1 + z_n) B_n(f, x)$, where $z_n$ is defined by

$$z_n = \begin{cases} 1 , & n \text{ is odd}, \\ 0 , & n \text{ is even} \end{cases}$$

Then,

$$B_n(1, x) = 1, \quad B_n(t, x) = x, \quad B_n(t^2, x) = x^2 + \frac{x - x^2}{n},$$

and the sequence $(P_n)$ satisfies the conditions (9.3.1)–(9.3.3). Hence, we have

$$F - \lim \| P_n(f, x) - f(x) \|_\infty = 0.$$

On the other hand, we get $P_n(f, 0) = (1 + z_n) f(0)$, since $B_n(f, 0) = f(0)$, and hence

$$\| P_n(f, x) - f(x) \|_\infty \geq |P_n(f, 0) - f(0)| = z_n |f(0)|.$$

We see that $(P_n)$ does not satisfy the classical Korovkin theorem, since $\limsup_{n \to \infty} z_n$ does not exist.

Our next result is an analogue of Theorem 9.2.2.

**Theorem 9.3.3.** *Let $(T_k)$ be a sequence of positive linear operators from $C_{2\pi}(R)$ into $C_{2\pi}(R)$. Then, for all $f \in C_{2\pi}(R)$*

$$F - \lim_{k \to \infty} \| T_k(f; x) - f(x) \|_{2\pi} = 0 \qquad (9.3.9)$$

*if and only if*

$$F - \lim_{k \to \infty} \| T_k(1; x) - 1 \|_{2\pi} = 0, \qquad (9.3.10)$$

$$F - \lim_{k \to \infty} \| T_k(\cos t; x) - \cos x \|_{2\pi} = 0, \qquad (9.3.11)$$

$$F - \lim_{k \to \infty} \| T_k(\sin t; x) - \sin x \|_{2\pi} = 0. \qquad (9.3.12)$$

*Proof.* Since each $f_0$, $f_1$, and $f_2$ belongs to $C_{2\pi}(\mathbb{R})$, the conditions (9.3.10)–(9.3.12) follow immediately from (9.3.9). Let the conditions (9.3.10)–(9.3.12) hold and $f \in C_{2\pi}(\mathbb{R})$.

Let $I$ be a closed subinterval of length $2\pi$ of $\mathbb{R}$. Fix $x \in I$. By the continuity of $f$ at $x$, it follows that for given $\varepsilon > 0$ there is a number $\delta > 0$ such that for all $t$

$$|f(t) - f(x)| < \varepsilon, \qquad (9.3.13)$$

whenever $|t - x| < \delta$. Since $f$ is bounded, it follows that

$$|f(t) - f(x)| \leq 2\|f\|_{2\pi},  \tag{9.3.14}$$

for all $t \in \mathbb{R}$. For all $t \in (x - \delta, 2\pi + x - \delta]$. Using (9.3.13) and (9.3.14), we obtain

$$|f(t) - f(x)| < \varepsilon + \frac{2\|f\|_{2\pi}}{\sin^2 \frac{\delta}{2}} \psi(t),  \tag{9.3.15}$$

where $\psi(t) = \sin^2[(t - x)/2]$. Since the function $f \in C_{2\pi}(\mathbb{R})$ is $2\pi$-periodic, the inequality (9.3.15) holds for $t \in \mathbb{R}$.

Now, operating $T_k(1; x)$ to this inequality, we obtain

$$|T_k(f; x) - f(x)| \leq [\varepsilon + |f(x)|]|T_k(1; x) - 1| + \varepsilon + \frac{\|f\|_{2\pi}}{\sin^2 \frac{\delta}{2}}[|T_k(1; x) - 1|$$

$$+|\cos x||T_k(\cos t; x) - \cos x| + |\sin x||T_k(\sin t; x) - \sin x|] \leq \varepsilon$$

$$+\left[\varepsilon + |f(x)| + \frac{\|f\|_{2\pi}}{\sin^2 \frac{\delta}{2}}\right]\{|T_k(1; x) - 1|$$

$$+|T_k(\cos t; x) - \cos x| + |T_k(\sin t; x) - \sin x|\}$$

Now, taking $\sup_{x \in I}$, we get

$$\|T_k(f; x) - f(x)\|_{2\pi} \leq \varepsilon + K \left(\|T_k(1; x) - 1\|_{2\pi}\right.$$

$$+ \|T_k(\cos t; x) - \cos x\|_{2\pi} + \|T_k(\sin t; x) - \sin x\|_{2\pi}\right),  \tag{9.3.16}$$

$$\text{where } K := \left\{\varepsilon + \|f\|_{2\pi} + \frac{\|f\|_{2\pi}}{\sin^2 \frac{\delta}{2}}\right\}.$$

Now replacing $T_k(\cdot, x)$ by $\frac{1}{m+1}\sum_{k=n}^{n+m} T_k(\cdot, x)$ in (9.3.17) on both sides and then taking the limit as $m \to \infty$ uniformly in $n$. Therefore, using conditions (9.3.10)–(9.3.12), we get

$$\lim_{m \to \infty} \left\|\frac{1}{m+1}\sum_{k=n}^{n+m} T_k(f, x) - f(x)\right\|_{2\pi} = 0 \text{ uniformly in } n,$$

i.e., the condition (9.3.9) is proved.

This completes the proof of the theorem.                                                    □

In the following example we see that Theorem 9.3.3 is stronger than Theorem 9.2.2.

**Theorem 9.3.4.** *For any $n \in \mathbb{N}$, denote by $S_n(f)$ the $n$-th partial sum of the Fourier series of $f$, i.e.,*

$$S_n(f)(x) = \frac{1}{2}a_0(f) + \sum_{k=1}^{n} a_k(f)\cos kx + b_k(f)\sin kx.$$

*For any $n \in \mathbb{N}$, write*

$$F_n(f) := \frac{1}{n+1}\sum_{k=0}^{n} S_k(f).$$

A standard calculation gives that for every $t \in \mathbb{R}$

$$
\begin{aligned}
F_n(f;x) &:= \frac{1}{2\pi}\int_{-\pi}^{\pi} f(t)\frac{1}{n+1}\sum_{k=0}^{n}\frac{\sin\frac{(2k+1)(x-t)}{2}}{\sin\frac{x-t}{2}}\,dt \\
&= \frac{1}{2\pi}\int_{-\pi}^{\pi} f(t)\frac{1}{n+1}\sum_{k=0}^{n}\frac{\sin^2\frac{(n+1)(x-t)}{2}}{\sin^2\frac{x-t}{2}}\,dt \\
&= \frac{1}{2\pi}\int_{-\pi}^{\pi} f(t)\varphi_n(x-t)\,dt,
\end{aligned}
$$

where

$$
\varphi_n(x) := 
\begin{cases}
\dfrac{\sin^2\frac{(n+1)(x-t)}{2}}{(n+1)\sin^2\frac{x-t}{2}} & ,\ x \text{ is not a multiple of } 2\pi, \\
n+1 & ,\ x \text{ is a multiple of } 2\pi.
\end{cases}
$$

The sequence $(\varphi_n)_{n\in\mathbb{N}}$ is a positive kernel which is called the *Fejér kernel,* and the corresponding operators $F_n$, $n \geq 1$ are called the *Fejér convolution operators.*

Note that the Theorem 9.2.2 is satisfied for the sequence $(F_n)$. In fact, we have for every $f \in C_{2\pi}(\mathbb{R})$, $F_n(f) \to f$, as $n \to \infty$.

Let $L_n : C_{2\pi}(\mathbb{R}) \to C_{2\pi}(\mathbb{R})$ be defined by

$$L_n(f;x) = (1+z_n)F_n(f;x), \tag{9.3.17}$$

where the sequence $z = (z_n)$ is defined as above. Now,

$$L_n(1;x) = 1,$$

$$L_n(\cos t; x) = \frac{n}{n+1}\cos x,$$

$$L_n(\sin t; x) = \frac{n}{n+1}\sin x$$

so that we have

$$F - \lim_{n\to\infty}\|L_n(1;x) - 1\|_{2\pi} = 0,$$

$$F - \lim_{n\to\infty} \|L_n(\cos t; x) - \cos x\|_{2\pi} = 0,$$

$$F - \lim_{n\to\infty} \|L_n(\sin t; x) - \sin x\|_{2\pi} = 0,$$

that is, the sequence $(L_n)$ satisfies the conditions (9.3.9)–(9.3.12). Hence by Theorem 9.3.3, we have

$$F\text{-} \lim_{n\to\infty} \|L_n(f) - f\|_{2\pi} = 0,$$

i.e., our theorem holds. But on the other hand, Theorem 9.2.2 does not hold for our operator defined by (9.3.17), since the sequence $(L_n)$ is not convergent.

Hence Theorem 9.3.3 is stronger than Theorem 9.2.2.

# Chapter 10
# Statistical Summability

## 10.1 Introduction

There is another notion of convergence known as the statistical convergence which was introduced by Fast [33] and Steinhaus [93] independently in 1951. In [66], Moricz mentioned that Henry Fast first time had heard about this concept from Steinhaus, but in fact it was Antoni Zygmund who proved theorems on the statistical convergence of Fourier series in the first edition of his book [101, pp. 181–188] where he used the term "almost convergence" in place of statistical convergence and at that time this idea was not recognized much. Since the term "almost convergence" was already in use (as described earlier in this book), Fast had to choose a different name for his concept and "statistical convergence" was most suitable. In this chapter we study statistical convergence and some of its variants and generalizations. Active researches were started after the paper of Fridy [37] and since then many of its generalizations and variants have appeared so far, e.g., [38, 62, 64, 70, 74, 76, 77], and so on.

## 10.2 Definitions and Notations

(i) Let $K \subseteq \mathbb{N}$. Then the *natural density* of $K$ is defined by (c.f. [24])

$$\delta(K) = \lim_{n \to \infty} \frac{1}{n} |\{k \leq n \; ; \; k \in K\}|,$$

where $|\{k \leq n : k \in K\}|$ denotes the number of elements of $K$ not exceeding $n$.

For example, the set of even integers has natural density $\frac{1}{2}$ and set of primes has natural density zero.

M. Mursaleen, *Applied Summability Methods*, SpringerBriefs in Mathematics,
DOI 10.1007/978-3-319-04609-9_10, © M. Mursaleen 2014

(ii) The number sequence $x$ is said to be *statistically convergent* to the number $L$ provided that for each $\epsilon > 0$,

$$\delta(K) = \lim_{n\to\infty} \frac{1}{n} |\{k \le n; |x_k - L| \ge \epsilon\}| = 0,$$

i.e.,

$$|x_k - L| < \epsilon \quad a.a.k. \tag{10.2.1}$$

In this case we write $\text{st} - \lim x_k = L$.

By the symbol st or $S$ we denote the set of all statistically convergent sequences and by $\text{st}_0$ or $S_0$ the set of all statistically null sequences.

*Remark 10.2.1.* Note that every convergent sequence is statistically convergent to the same number, so that statistical convergence is a natural generalization of the usual convergence of sequences. The sequence which converges statistically need not be convergent and also need not be bounded.

*Example 10.2.2.* Let $x = (x_k)$ be defined by

$$x_k = \begin{cases} k & , k \text{ is a square,} \\ 0 & , \text{otherwise.} \end{cases}$$

Then $|\{k \le n : x_k \ne 0\}| \le \sqrt{n}$. Therefore, $\text{st} - \lim x_k = 0$. Note that we could have assigned any values whatsoever to $x_k$ when $k$ is a square, and we could still have $\text{st} - \lim x_k = 0$. But $x$ is neither convergent nor bounded.

It is clear that if the inequality in (10.2.1) holds for all but finitely many $k$, then $\lim x_k = L$. It follows that $\lim x_k = L$ implies $\text{st} - \lim x_k = L$ so statistical convergence may be considered as a regular summability method. This was observed by Schoenberg [90] along with the fact that the statistical limit is a linear functional on some sequence space. Salat [87] proved that the set of bounded statistically convergent (real) sequences is a closed subspace of the space of bounded sequences.

In most convergence theories it is desirable to have a criterion that can be used to verify convergence without using the value of the limit. For this purpose we introduce the analogue of the Cauchy convergence criterion [37].

(iii) The number sequence $x$ is said to be *statistically Cauchy sequence* provided that for every $\epsilon > 0$ there exists a number $N(= N(\epsilon))$ such that

$$|x_k - x_N| < \epsilon \quad a.a.k, \tag{10.2.2}$$

i.e.,

$$\lim_{n\to\infty} \frac{1}{n} |\{k \le n : |x_n - x_N| \ge \epsilon\}| = 0.$$

In order to prove the equivalence of Definitions given in Parts (i) and (ii) of Sect. 10.2 we shall find it helpful to use a third (equivalent). This property states that for almost all $k$, the values $x_k$ coincide with those of a convergent sequence.

## 10.3   Results on Statistical Convergence

**Theorem 10.3.1.** *The following statements are equivalent:*

*(i)  x is a statistically convergent sequence;*
*(ii)  x is a statistically Cauchy sequence;*
*(iii)  x is a sequence for which there is a convergent sequence y such that $x_k = y_k$ a.a.k.*

As an immediate consequence of Theorem 10.3.1 we have the following result.

**Corollary 10.3.2.** *If x is a sequence such that* st-lim $x_k = L$, *then x has a subsequence y such that* lim $y_k = L$.

Schoenberg [90, Lemma 4] proved that the Cesàro mean of order 1 sums every bounded statistically convergent sequence. This raises the question of whether the $C_1$ method includes the statistical convergence method irrespective of boundedness. The answer is negative, a fortiori, as we shall see in the next theorem. But first we give a useful lemma.

**Lemma 10.3.3.** *If t is a number sequence such that $t_k \neq 0$ for infinitely many k, then there is a sequence x such that $x_k = 0$ a.a.k. and $\sum_{k=1}^{\infty} t_k x_k = \infty$.*

*Proof.* Choose an increasing sequence of positive integers $\{m(k)\}_{k=1}^{\infty}$ such that for each $k$,

$$m(k) > k^2 \text{ and } t_{m(k)} \neq 0.$$

Define $x$ by $x_{m(k)} = 1/t_{m(k)}$ and $x_k = 0$ otherwise. Then $x_k = 0$ a.a.k and $\sum_{k=1}^{\infty} t_k x_k = \sum_{k=1}^{\infty} t_{m(k)} x_{m(k)} = \infty$.                                     □

**Theorem 10.3.4.** *No matrix summability method can include the method of statistical convergence.*

*Proof.* The preceding Lemma 10.3.3 shows that in order for a matrix to include statistical convergence it would have to be row-finite. Let $A$ be an arbitrary row-finite matrix and choose a nonzero entry, say $a_{n(1),k'(1)} \neq 0$. Then choose $k(1) \geq k'(1)$ so that

$$a_{n(1),k(1)} \neq 0 \text{ and } a_{n(1),k} = 0 \text{ if } k > k(1).$$

Now select increasing sequences of row and column indices such that for each $m$,

$$a_{n(m),k(m)} \neq 0, \ k(m) \geq m^2, \ \text{and} \ a_{n(m),k} = 0 \ \text{if} \ k > k(m).$$

Define the sequence $x = (x_k)$ as follows:

$$x_k = \begin{cases} \frac{1}{a_{n(1),k(1)}}, \cdots & , k = k(1), \\ \frac{1}{a_{n(m),k(m)}} \left[ m - \sum_{i=1}^{m-1} a_{n(m),k(i)} x_{k(i)} \right], \cdots , & k = k(m), \\ 0 & , \text{otherwise}. \end{cases}$$

Then $x$ is not $A$-summable because $(Ax)_{n(m)} = m$; also, $k(m) \geq m^2$ implies that $|\{k \leq n : x_k \neq 0\}| \leq \sqrt{n}$, so $x_k = 0$ $a.a.k$. Thus st $-\lim x_k = 0$, we conclude that $A$ does not include statistical convergence.                                         □

*Remark 10.3.5.* By definition, the method of statistical convergence cannot sum any periodic sequence such as $\{(-1)^k\}$. Therefore, statistical convergence does not include most of the classical summability methods. When combined with Theorem 10.3.4 this suggests that perhaps statistical convergence cannot be compared to any nontrivial matrix method. The following example shows that is not the case.

*Example 10.3.6.* Define the matrix $A$ by

$$a_{nk} = \begin{cases} 1 & , k = n \text{ and } n \text{ is not a square}, \\ 1/2 & , n = m^2 \text{ and } k = n \text{ or } k = (m-1)^2, \ m \in \mathbb{N} \\ 0 & , \text{otherwise}. \end{cases}$$

Then for any sequence $x$ we have

$$(Ax)_n = \begin{cases} x_1/2 & , n = 1, \\ [x_{(m-1)^2} + x_{m^2}]/2 & , n = m^2 \text{ for } m = 2,3,\ldots \\ x_n & , n \text{ is not a square}. \end{cases}$$

Thus $A$ is obviously a regular triangle. To see that $A$ is included by statistical convergence suppose $\lim_{n \to \infty}(Ax)_n = L$. Then $\lim_{\substack{n \to \infty \\ n \neq m}} x_n = L$ and obviously $|\{k \leq n : (Ax)_n \neq x_n\}| \leq \sqrt{n}$, so by Theorem 10.3.1, st-$\lim x_k = L$. To see that $A$ is not equivalent to ordinary convergence consider the sequence $x = (x_k)$ given by

$$x_k = \begin{cases} (-1)^m & , k = m^2 \text{ for } m = 1,2,\ldots, \\ 0 & , k \text{ is not a square}. \end{cases}$$

Then $(Ax)_n = 0$ for $n > 1$, but $x$ is not convergent.

*Remark 10.3.7.* We know that every subsequence of a convergent sequence is convergent, but this is no longer true in case of statistical convergence, i.e., a statistically convergent sequence may have a subsequence which is not statistically convergent. Consider the statistically convergent sequence $x = (x_k)$ as defined in Example 10.2.2. Now, consider the subsequence $(x_{k^2})$ of $(x_k)$. It is clear that the subsequence $(x_{k^2})$ of the statistically convergent sequence $(x_k)$ is not statistically convergent.

In this direction we state the following important result given by Salat [87], which tells about the structure of a statistically convergent sequence.

**Theorem 10.3.8.** *A sequence $x = (x_k)$ is statistically convergent to $L$ if and only if there exists a set $K = \{k_1 < k_2 < \cdots < k_n < \cdots\} \subseteq \mathbb{N}$ such that $\delta(K) = 1$ and $\lim x_{k_n} = L$.*

*Proof.* Suppose that there exists a set $K = \{k_1 < k_2 < \cdots < k_n < \cdots\} \subseteq \mathbb{N}$ such that $\delta(K) = 1$ and $\lim_{n \to \infty} x_{k_n} = L$. Then there is a positive integer $N$ such that for $n > N$,

$$|x_{k_n} - L| < \epsilon. \tag{10.3.1}$$

Put $K_\epsilon := \{n \in \mathbb{N} : |x_n - L| \geq \epsilon\}$ and $K' = \{k_{N+1}, k_{N+2}, \ldots\}$. Then $\delta(K') = 1$ and $K_\epsilon \subseteq \mathbb{N} - K'$ which implies that $\delta(K_\epsilon) = 0$. Hence $x = (x_k)$ is statistically convergent to $L$.

Conversely, let $x = (x_k)$ be statistically convergent to $L$. For $r = 1, 2, 3, \ldots$, put $K_r := \{n \in \mathbb{N} : |x_n - L| \geq 1/r\}$ and $M_r := \{n \in \mathbb{N} : |x_n - L| < 1/r\}$. Then $\delta(K_r) = 0$ and

$$M_1 \supset M_2 \supset \cdots M_i \supset M_{i+1} \supset \cdots \tag{10.3.2}$$

and

$$\delta(M_r) = 1, \; r = 1, 2, 3, \ldots \tag{10.3.3}$$

Now we have to show that for $n \in M_r$, $(x_{k_n})$ is convergent to $L$. Suppose that $(x_{k_n})$ is not convergent to $L$. Therefore there is $\epsilon > 0$ such that $|x_{k_n} - L| \geq \epsilon$ for infinitely many terms. Let $M_\epsilon := \{n \in \mathbb{N} : |x_{k_n} - L| < \epsilon\}$ and $\epsilon > 1/r$ ($r = 1, 2, 3, \ldots$). Then

$$\delta(M_\epsilon) = 0, \tag{10.3.4}$$

and by (10.3.2), $M_r \subset M_\epsilon$. Hence $\delta(M_r) = 0$, which contradicts (10.3.3), and therefore $(x_{k_n})$ is convergent to $L$.

This completes the proof of the theorem. □

## 10.4   Strong Cesàro Summability

Here, we define strong $p$-Cesàro summability and find its connection with statistical convergence.

**Definition 10.4.1.** Let $p \in \mathbb{R}$, $0 < p < \infty$. A sequence $x = (x_k)$ is said to be *strongly $p$-Cesàro summable* to the limit $L$ if $\sum_{k=1}^{n} |x_k - L|^p / n \to 0$, as $n \to \infty$. In this case, we write $x_k \to L[C,1]_p$.

The following result provides the relationship between strongly $p$-Cesàro summability and statistical convergence (c.f. [25, 101]).

**Theorem 10.4.2.** *If a sequence is strongly $p$-Cesàro summable to $L$, then it is statistically convergent to $L$. If a bounded sequence is statistically convergent to $L$, then it is strongly $p$-Cesàro summable to $L$.*

*Proof.* Let $x = (x_k)$ be any strongly $p$-Cesàro summable sequence to $L$. Then, for a given $\epsilon > 0$, we have

$$\sum_{k=1}^{n} |x_k - L|^p \geq \left| \{k \leq n : |x_k - L|^p \geq \epsilon\} \right| \epsilon^p.$$

It follows that if $x$ is strongly $p$-Cesàro summable to $L$, then $x$ is statistically convergent to $L$.

Now suppose that $x$ is bounded and statistically convergent to $L$ and put $K = \|x\|_\infty + |L|$. Let $\epsilon > 0$ be given and select $N_\epsilon$ such that

$$\frac{1}{n} \left| k \leq n : |x_k - L|^p \geq \left(\frac{\epsilon}{2}\right)^{1/p} \right| < \frac{\epsilon}{2K^p},$$

for all $n > N_\epsilon$ and set $L_n = \left\{ k \leq n : |x_k - L| \geq (\epsilon/2)^{1/p} \right\}$. Now for $n > N_\epsilon$ we have that

$$\frac{1}{n} \sum_{k=1}^{n} |x_k - L|^p = \frac{1}{n} \left\{ \sum_{k \in L_n} |x_k - L|^p + \sum_{k \notin L_n, k \leq n} |x_k - L|^p \right\}$$

$$< \frac{n\epsilon}{2nk^p} k^p + \frac{n\epsilon}{2n}$$

$$= \frac{\epsilon}{2} + \frac{\epsilon}{2} = \epsilon.$$

Hence, $x$ is strongly $p$-Cesàro summable to $L$.                                       □

The following important result is due to Connor [25].

**Theorem 10.4.3 (Decomposition Theorem).** *If $x \in \omega$ is strongly $p$-Cesàro summable or statistically convergent to $L$, then there is a convergent sequence $y$ and a statistically null sequence $z$ such that $y$ is convergent to $L$, $x = y + z$ and $\lim_n n^{-1} |\{k \leq n : z_k \neq 0\}| = 0$. Moreover, if $x$ is bounded then $\|z\|_\infty \leq \|x\|_\infty + |L|$.*

*Proof.* We apply Theorem 10.4.2. Let $N_0 = 0$ and select an increasing sequence of positive integers $N_1 < N_2 < N_3 < \ldots$ such that if $n > N_j$ we have that

$$n^{-1}\big|\{k \le n : |x_k - L| \ge j^{-1}\}\big| < j^{-1}.$$

Now define $y$ and $z$ as follows: if $N_0 < k < N_1$ set $z_k = 0$ and $y_k = x_k$. Now suppose that $j \ge 1$ and that $N_j < k \le N_{j+1}$. If $|x_k - L| < j^{-1}$ we set $y_k = x_k$ and $z_k = 0$ and if $|x_k - L| \ge j^{-1}$ we set $y_k = L$ and $z_k = x_k - L$. It is clear from our construction that $x = y + z$ and that $\|z\|_\infty \le \|x\|_\infty + |L|$ if $x$ is bounded.

We first show that $\lim_k y_k = L$. Let $\epsilon > 0$ and pick $j$ such that $\epsilon > j^{-1}$. Observe that for $k > N_j$ we have that $|y_k - L| < \epsilon$ since $|y_k - L| = |x_k - L| < \epsilon$ if $|x_k - L| < j^{-1}$ and $|y_k - L| = |L - L| = 0$ if $|x_k - L| > j^{-1}$. Since $\epsilon$ was arbitrary, we have established the claim.

Next we prove that $z$ is statistically null. Note that it suffices to show that $\lim_n n^{-1}|\{k \le n : z_k \ne 0\}| = 0$, which follows by observing that $|\{k \le n : z_k \ne 0\}| \ge |\{k \le n : |z_k| \ge \epsilon\}|$ for any natural number $n$ and $\epsilon > 0$.

We now show that if $\delta > 0$ and $j \in N$ such that $j^{-1} < \delta$, then $|\{k \le n : z_k \ne 0\}| < \delta$ for all $n > N_j$. Recall from the construction that if $N_j < k \le N_{j+1}$, then $z_k \ne 0$ only if $|x_k - L| > j^{-1}$. It follows that if $N_\ell < k \le N_{\ell+1}$, then

$$\{k \le n : z_k \ne 0\} \subseteq \{k \le n : |x_k - L| > \ell^{-1}\}.$$

Consequently, if $N_\ell < n \le N_{\ell+1}$ and $\ell > j$, then

$$n^{-1}\big|\{k \le n : z_k \ne 0\}\big| \le n^{-1}\big|\{k \le n : |x_k - L| > \ell^{-1}\}\big| < \ell^{-1} < j^{-1} < \delta.$$

This completes the proof of the theorem.                                    □

We deduce the following corollary.

**Corollary 10.4.4.** *Let $x \in \omega$. If $x$ is strongly $p$-Cesàro summable to $L$ or statistically convergent to $L$, then $x$ has a subsequence which converges to $L$.*

## 10.5  Application to Fourier Series

Let $f : \mathbb{T} \to \mathbb{C}$ be a Lebesgue integrable function on the torus $\mathbb{T} := [-\pi, \pi)$, i.e., $f \in L^1(\mathbb{T})$. The Fourier series of $f$ is defined by

$$f(x) \sim \sum_{j \in \mathbb{Z}} \hat{f}(j)e^{ijx}, \quad x \in \mathbb{T}, \tag{10.5.1}$$

where the Fourier coefficients $\hat{f}(j)$ are defined by

$$\hat{f}(j) := \frac{1}{2\pi} \int_{\mathbb{T}} f(t)e^{-ijt}\, dt, \quad j \in \mathbb{Z}. \tag{10.5.2}$$

The symmetric partial sums of the series in (10.5.1) are defined by

$$s_k(f;x) := \sum_{|j|\le k} \hat{f}(j)e^{ijx}, \quad x \in \mathbb{T}, \ k \in \mathbb{N}. \tag{10.5.3}$$

The conjugate series to the Fourier series in (10.5.1) is defined by [101, vol. I, p. 49]

$$\sum_{j\in\mathbb{Z}} (-i \ \text{sgn} \ j) \hat{f}(j)e^{ijx}. \tag{10.5.4}$$

Clearly, it follows from (10.5.1) and (10.5.4) that

$$\sum_{j\in\mathbb{Z}} \hat{f}(j)e^{ijx} + i \sum_{j\in\mathbb{Z}} (-i \ \text{sgn} \ j) \hat{f}(j)e^{ijx} = 1 + 2\sum_{j=1}^{\infty} \hat{f}(j)e^{ijx},$$

and the power series

$$1 + 2\sum_{j=1}^{\infty} \hat{f}(j)e^{ijx}, \quad \text{where } z := re^{ix}, \ 0 \le r < 1,$$

is analytic on the open unit disk $|z| < 1$, due to the fact that

$$|\hat{f}(j)| \le \frac{1}{2\pi} \int_{\pi} |f(t)| \, dt, \quad j \in \mathbb{Z}.$$

The conjugate function $\hat{f}$ of a function $f \in L^1(\mathbb{T})$ is defined by

$$\begin{aligned} \hat{f}(x) :=& -\lim_{\varepsilon\downarrow 0} \frac{1}{\pi} \int_{\varepsilon\le|t|\le\pi} \frac{f(x+t)}{2\tan\frac{t}{2}} dt \\ =& \lim_{\varepsilon\downarrow 0} \frac{1}{\pi} \int_{\varepsilon}^{\pi} \frac{f(x-t) - f(x+t)}{2\tan\frac{t}{2}} dt \end{aligned} \tag{10.5.5}$$

in the "principal value" sense, and that $\hat{f}(x)$ exists at almost every $x \in \mathbb{T}$.

We have the following results [100] (c.f. [67, Theorem 2.1 (ii)]).

**Theorem 10.5.1.** *If $f \in L^1(\mathbb{T})$, then for any $p > 0$ its Fourier series is strongly p-Cesàro summable to $f(x)$ at almost every $x \in \mathbb{T}$. Furthermore, its conjugate series (10.5.4) is strongly p-Cesàro summable for any $p > 0$ to the conjugate function $\hat{f}(x)$ defined in (10.5.5) at almost every $x \in \mathbb{T}$.*

The above result together with Theorem 10.4.2 implies the following useful result.

**Theorem 10.5.2.** *If $f \in L^1(\mathbb{T})$, then its Fourier series is statistically convergent to $f(x)$ at almost every $x \in \mathbb{T}$. Furthermore, its conjugate series (10.5.4) is statistically convergent to the conjugate function $\hat{f}(x)$ defined in (10.5.5) at almost every $x \in \mathbb{T}$.*

## 10.6 A-Statistical Convergence

In this section, we study the notion of $A$-density and $A$-statistical convergence. Following Freedman and Sember [36], Kolk [54] introduced the notion of $A$-statistical convergence by taking an arbitrary nonnegative regular matrix $A$ in place of Cesàro matrix $C_1$ in the definition of statistical convergence.

**Definition 10.6.1.** Let $K = \{k_i\}$ be an index set and let $\varphi^K = (\varphi_j^K)$ with

$$\varphi_j^K = \begin{cases} 1, & j \in K, \\ 0, & \text{otherwise.} \end{cases}$$

For a nonnegative regular matrix $A$, if $A\varphi^K \in c$ (the space of convergent sequences), then $\delta_A(K) = \lim_{n\to\infty} A_n\varphi^K$ is called the $A$-*density* of $K$, thus

$$\delta_A(K) = \lim_{n\to\infty} \sum_{k\in K} a_{nk} = \lim_{n\to\infty} \sum_i a_{n,k_i}.$$

**Definition 10.6.2.** A sequence $x = (x_k)$ is said to be $A$-*statistically convergent* to the number $L$ if $\delta_A(K_\epsilon) = 0$ for every $\epsilon > 0$, where $K_\epsilon = \{k : |x_k - L| \geq \epsilon\}$. In this case, we write $\text{st}_A - \lim x_k = L$. By the symbol $\text{st}_A$ we denote the set of all $A$-statistically convergent sequences and by $\text{st}_A^0$ the set of all $A$-statistically null sequences.

It should be noted that $A$-statistical convergence is defined only for a nonnegative matrix $A$.

**Definition 10.6.3.** A matrix $A = (a_{nk})$ is called *uniformly regular* if it satisfies the following conditions:

(i) $\sup_{n\in\mathbb{N}} \sum_{k=0}^{\infty} |a_{nk}| < \infty$;
(ii) $\lim_{n\to\infty} \sum_{k=0}^{\infty} a_{nk} = 1$;
(iii) $\lim_{n\to\infty} \sup_{k\in\mathbb{N}} |a_{nk}| = 0$.

Agnew [3] has proved the following theorem:

**Theorem 10.6.4 (Agnew's Theorem).** *If a matrix $A = (a_{nk})$ satisfies the condition $\lim_{n\to\infty} \sup_{k\in\mathbb{N}} |a_{nk}| = 0$ and $\sum_{k=0}^{\infty} |a_{nk}| < \infty$ for all $n \in \mathbb{N}$, then there exists a divergent sequence of 0s and 1s which is $A$-summable to 0 or, equivalently, if $A$*

*satisfies the assumptions of Agnew's theorem, then there exists an infinite index set* $K$ *with* $\delta_A(K) = 0$.

For a uniformly regular matrix $A$ and an infinite index set $K = \{k_i\}$ the submatrix $(a_{n,k_i})$ of $A$ obviously satisfies the assumptions of Agnew's theorem. Therefore, we obtain the following result:

**Theorem 10.6.5.** *If the matrix* $A$ *is uniformly regular then every infinite index set contains an infinite subset* $K$ *with* $\delta_A(K) = 0$.

We have the following important characterization of $A$-statistical convergence, proved for $A = C_1$ by Fridy [37] and for an arbitrary nonnegative regular $A$ by Kolk [54], which is an analogue of Theorem 10.3.8.

**Theorem 10.6.6.** *A sequence* $x = (x_k)$ *converges* $A$*-statistically to* $L$ *if and only if there exists an infinite index set* $K = \{k_i\}$ *so that the subsequence* $(x_{k_i})$ *converges to* $L$ *and* $\delta_A(\mathbb{N} \setminus K) = 0$ *(and hence* $\delta_A(K) = 1$*).*

Note that Theorem 10.6.6 together with Theorem 10.6.5 shows that for a uniformly regular matrix $A$ the $A$-statistical convergence is strictly stronger than convergence.


## 10.7   Statistical $A$-Summability

Recently, the idea of statistical $(C, 1)$-summability was introduced in [63], of statistical $(H, 1)$-summability in [63] by Moricz, and of statistical $(\bar{N}, p)$-summability by Moricz and Orhan [68]. In this section we generalize these statistical summability methods by defining the statistical $A$-summability for a nonnegative regular matrix $A$ which is due to Edely and Mursaleen [30] and find its relationship with $A$-statistical convergence. Statistical $A$-summability for double sequences is studied in [15].

**Definition 10.7.1.** Let $A = (a_{ik})$ be a nonnegative regular matrix and $x = (x_k)$ be a sequence. We say that $x$ is *statistically* $A$*-summable* to $\ell$ if for every $\epsilon > 0$, $\delta(\{i \leq n : |y_i - \ell| \geq \epsilon\}) = 0$, i.e.,

$$\lim_{n \to \infty} \frac{1}{n} |\{i \leq n : |y_i - \ell| \geq \epsilon\}| = 0,$$

where $y_i = A_i(x)$. Thus $x$ is statistically $A$-summable to $\ell$ if and only if $Ax$ is statistically convergent to $\ell$. In this case we write $\ell = (A)_{st} - \lim x \, (= st - \lim Ax)$. By $(A)_{st}$, we denote the set of all statistically $A$-summable sequences.

*Remark 10.7.2.* We have the following particular cases:

(i) If we take $A = (a_{ik})$ defined by

$$a_{ik} = \begin{cases} \frac{1}{i+1} , 0 \le k \le i, \\ 0 , \text{ otherwise}, \end{cases}$$

then the statistical $A$-summability is reduced to the statistical $(C, 1)$-summability due to *Moricz* [63].

(ii) If we take $A = (a_{ik})$ defined by

$$a_{ik} = \begin{cases} \frac{p_k}{P_i} , 0 \le k \le i, \\ 0 , \text{ otherwise}, \end{cases}$$

then the statistical $A$-summability is reduced to the statistical $(\bar{N}, p)$-summability due to *Moricz and Orhan* [68], where $p = (p_k)$ is a sequence of nonnegative numbers such that $p_0 > 0$ and

$$\lim_{i \to \infty} P_i = \lim_{i \to \infty} \sum_{k=0}^{i} p_k = \infty.$$

(iii) If we take $A = (a_{ik})$ defined by

$$a_{ik} = \begin{cases} \frac{1}{k l_i} , 0 \le k \le i, \\ 0 , \text{ otherwise}, \end{cases}$$

where $l_i = \sum_{k=0}^{i} 1/(k + 1)$, then the statistical $A$-summability is reduced to the statistical $(H, 1)$-summability due to *Moricz* [65].

(iv) If we take $A = (a_{nk})$ defined by

$$a_{nk} = \begin{cases} \frac{1}{\lambda_n} , k \in I_n = [n - \lambda_n + 1, n], \\ 0 , k \notin I_n, \end{cases}$$

then the statistical $A$-summability is reduced to the statistical $\lambda$-summability due to *Mursaleen and Alotaibi* [71], where $\lambda = (\lambda_n)$ is a nondecreasing sequence of positive numbers tending to $\infty$ such that $\lambda_{n+1} \le \lambda_n + 1, \lambda_1 = 0$.

(v) If we take $A = (a_{nk})$ defined by

$$a_{nk} = \begin{cases} \frac{1}{h_r} , k \in I_r = (k_{r-1}, k_r], \\ 0 , k \notin I_r, \end{cases}$$

then the statistical $A$-summability is reduced to the statistical lacunary summability due to Mursaleen and Alotaibi [72], where $\theta = (k_r)$ is a lacunary sequence such that $k_0 = 0$ and $h_r = k_r - k_{r-1} \to \infty$ as $r \to \infty$.

We give the relation between statistical $A$-summability and $A$-statistical convergence.

**Theorem 10.7.3.** *If a sequence is bounded and $A$-statistically convergent to $l$, then it is $A$-summable to $l$ and hence statistically $A$-summable to $l$ but not conversely.*

*Proof.* Let $x$ be bounded and $A$-statistically convergent to $l$, and $K_\epsilon = \{k \leq n : |x_k - l| \geq \epsilon\}$. Then

$$|A_n(x) - l| \leq \left| \sum_{k \notin K_\epsilon} a_{nk}(x_k - l) \right| + \left| \sum_{k \in K_\epsilon} a_{nk}(x_k - l) \right|$$

$$\leq \epsilon \sum_{k \notin K_\epsilon} a_{nk} + \sup_{k \in \mathbb{N}} |(x_k - l)| \sum_{k \in K_\epsilon} a_{nk}.$$

By using the definition of $A$-statistical convergence and the conditions of regularity of $A$, we get $|A_n(x) - l| \to 0$, as $n \to \infty$, since $\epsilon$ is arbitrary and hence $st - \lim |A_n(x) - l| = 0$.

To see that the converse does not hold, we construct the following examples:

(i) Let $A = (a_{nk})$ be the Cesàro matrix, i.e.,

$$a_{nk} = \begin{cases} \frac{1}{n+1} , & 0 \leq n \leq k, \\ 0 , & \text{otherwise}, \end{cases}$$

and let

$$x_k = \begin{cases} 1 , & k \text{ is odd}, \\ 0 , & k \text{ is even}. \end{cases}$$

Then $x$ is $A$-summable to $1/2$ (and hence statistically $A$-summable to $1/2$) but not $A$-statistically convergent.

(ii) Take $x = (x_k)$ as above and let $A = (a_{nk})$ be defined by

$$a_{nk} = \begin{cases} 1/2 , & n \text{ is non-square and } k = n^2, n^2 + 1, \\ 1 , & n \text{ is a square and } k = n^2, \\ 0 , & \text{otherwise}. \end{cases}$$

Then, we have

$$\sum_{k=1}^{\infty} a_{nk} x_k = \begin{cases} 1/2 \ , & n \ \text{is a nonsquare,} \\ 0 \ \ , & n \ \text{is even square,} \\ 1 \ \ , & \text{otherwise.} \end{cases}$$

We see that $x$ is not $A$-summable and hence it is not $A$-statistically convergent but

$$\lim_{n \to \infty} \frac{1}{n} \left| \left\{ i \le n : \left| y_i - \frac{1}{2} \right| \ge \epsilon \right\} \right| = 0,$$

i.e., $x$ is statistically $A$-summable to $1/2$.                                          □

# Chapter 11
# Statistical Approximation

## 11.1 Introduction

In the last chapter we discussed statistical summability and its various
generalizations and variants, e.g., lacunary statistical convergence, $\lambda$-statistical
convergence, $A$-statistical convergence, statistical summability $(C, 1)$, and
statistical $A$-summability. In this chapter, we demonstrate some applications of these
summability methods in proving Korovkin-type approximation theorems. Such a
method was first used by Gadjiev and Orhan [39] in which the statistical version of
Korovkin approximation was proved by using the test functions $1$, $x$, and $x^2$. Since
then a large amount of work has been done by applying statistical convergence and
its variants, e.g., [61,71–73,75,92] for different set of test functions. In this chapter
we present few of them and demonstrate the importance of using these new methods
of summability.

## 11.2 Application of Statistical Summability $(C, 1)$

In this section, we use the notion of statistical summability $(C, 1)$ to prove the
Korovkin-type approximation theorem by using the test functions $1, e^{-x}, e^{-2x}$. We
apply the classical Baskakov operator to construct an example in support of this
result.

For a sequence $x = (x_k)$, let us write $t_n = \frac{1}{n+1} \sum_{k=0}^{n} x_k$. We say that a sequence
$x = (x_k)$ is *statistically summable* $(C, 1)$ if $\text{st} - \lim_{n\to\infty} t_n = L$. In this case we
write $L = C_1(\text{st}) - \lim x$.

First we demonstrate through the following example that the statistical summability $(C, 1)$ is stronger than both ordinary convergence as well as statistical
convergence.

M. Mursaleen, *Applied Summability Methods*, SpringerBriefs in Mathematics,
DOI 10.1007/978-3-319-04609-9_11, © M. Mursaleen 2014

*Example 11.2.1.* Define the sequence $x = (x_k)$ by

$$x_k = \begin{cases} 1 & , k = m^2 - m, \ m^2 - m + 1, \dots, m^2 - 1, \\ -m & , k = m^2, \ m = 2, 3, 4, \dots, \\ 0 & , \text{ otherwise.} \end{cases} \qquad (11.2.1)$$

Then

$$t_n = \frac{1}{n+1} \sum_{k=0}^{n} x_k$$

$$= \begin{cases} \frac{s+1}{n+1} & , n = m^2 - m + s, \ s = 0, 1, 2, \dots, m - 1; \ m = 2, 3, \dots, \\ 0 & , \text{ otherwise.} \end{cases}$$

We easily see that $t_n \to 0$, as $n \to \infty$ and hence $st - \lim_{n \to \infty} t_n = 0$, i.e., $x = (x_k)$ is statistically summable $(C, 1)$ to 0. On the other hand $st - \liminf_{k \to \infty} x_k = 0$ and $st - \limsup_{k \to \infty} x_k = 1$, since the sequence $(m^2)_{m=2}^{\infty}$ is statistically convergent to 0. Hence, $x = (x_k)$ is not statistically convergent.

Let $C(I)$ be the Banach space with the uniform norm $\| \cdot \|_\infty$ of all real-valued continuous functions on $I = [0, \infty)$; provided that $\lim_{x \to \infty} f(x)$ is finite.

Boyanov and Veselinov [19] have proved the following theorem on $C[0, \infty)$ by using the test functions $1, e^{-x}, e^{-2x}$.

**Theorem 11.2.2.** *Let $(T_k)$ be a sequence of positive linear operators from $C(I)$ into $C(I)$. Then for all $f \in C(I)$*

$$\lim_{k \to \infty} \| T_k(f; x) - f(x) \|_\infty = 0$$

*if and only if*

$$\lim_{k \to \infty} \| T_k(1; x) - 1 \|_\infty = 0,$$

$$\lim_{k \to \infty} \| T_k(e^{-s}; x) - e^{-x} \|_\infty = 0,$$

$$\lim_{k \to \infty} \| T_k(e^{-2s}; x) - e^{-2x} \|_\infty = 0.$$

Now we prove the following result by using the notion of statistical summability $(C, 1)$.

**Theorem 11.2.3.** *Let $(T_k)$ be a sequence of positive linear operators from $C(I)$ into $C(I)$. Then for all $f \in C(I)$*

$$C_1(\text{st}) - \lim_{k \to \infty} \| T_k(f; x) - f(x) \|_\infty = 0 \qquad (11.2.2)$$

*if and only if*

$$C_1(\mathrm{st}) - \lim_{k \to \infty} \| T_k(1; x) - 1 \|_\infty = 0, \tag{11.2.3}$$

$$C_1(\mathrm{st}) - \lim_{k \to \infty} \| T_k(e^{-s}; x) - e^{-x} \|_\infty = 0, \tag{11.2.4}$$

$$C_1(\mathrm{st}) - \lim_{k \to \infty} \| T_k(e^{-2s}; x) - e^{-2x} \|_\infty = 0. \tag{11.2.5}$$

*Proof.* Since each $1, e^{-x}, e^{-2x}$ belongs to $C(I)$, conditions (11.2.3)–(11.2.5) follow immediately from (11.2.2). Let $f \in C(I)$. Then there exists a constant $M > 0$ such that $|f(x)| \le M$ for $x \in I$. Therefore,

$$|f(s) - f(x)| \le 2M, \quad -\infty < s, x < \infty. \tag{11.2.6}$$

Also, for a given $\varepsilon > 0$ there is a $\delta > 0$ such that

$$|f(s) - f(x)| < \varepsilon, \tag{11.2.7}$$

whenever $|e^{-s} - e^{-x}| < \delta$ for all $x \in I$.
   Using (11.2.6), (11.2.7), putting $\psi_1 = \psi_1(s, x) = (e^{-s} - e^{-x})^2$, we get

$$|f(s) - f(x)| < \varepsilon + \frac{2M}{\delta^2}(\psi_1), \ \forall \ |s - x| < \delta.$$

This is,

$$-\varepsilon - \frac{2M}{\delta^2}(\psi_1) < f(s) - f(x) < \varepsilon + \frac{2M}{\delta^2}(\psi_1).$$

Now, operating $T_k(1; x)$ to this inequality since $T_k(f; x)$ is monotone and linear, we obtain

$$T_k(1; x)\left[ -\varepsilon - \frac{2M}{\delta^2}(\psi_1) \right] < T_k(1; x)(f(s) - f(x))$$

$$< T_k(1; x)\left[ \varepsilon + \frac{2M}{\delta^2}(\psi_1) \right].$$

Note that $x$ is fixed and so $f(x)$ is a constant number. Therefore,

$$-\varepsilon T_k(1; x) - \frac{2M}{\delta^2} T_{j,k}(\psi_1; x) < T_k(f; x) - f(x) T_k(1; x)$$

$$< \varepsilon T_k(1; x) + \frac{2M}{\delta^2} T_k(\psi_1; x). \tag{11.2.8}$$

But

$$T_k(f;x) - f(x) = T_k(f;x) - f(x)T_k(1;x) + f(x)T_k(1;x) - f(x)$$
$$= [T_k(f;x) - f(x)T_k(1;x)] + f(x)[T_k(1;x) - 1].$$

(11.2.9)

Using (11.2.8) and (11.2.9), we have

$$T_k(f;x) - f(x) < \varepsilon T_k(1;x) + \frac{2M}{\delta^2}T_k(\psi_1;x) + f(x)[T_k(1;x) - 1].$$

(11.2.10)

Now

$$T_k(\psi_1;x) = T_k[(e^{-s} - e^{-x})^2;x] = T_k(e^{-2s} - 2e^{-s}e^{-x} + e^{-2x};x)$$
$$= T_k(e^{-2s};x) - 2e^{-x}T_k(e^{-s};x) + (e^{-2x})T_k(1;x)$$
$$= [T_k(e^{-2s};x) - e^{-2x}] - 2e^{-x}[T_k(e^{-s};x) - e^{-x}] + e^{-2x}[T_k(1;x) - 1].$$

Using (11.2.10), we obtain

$$T_k(f;x) - f(x) < \varepsilon T_k(1;x) + \frac{2M}{\delta^2}\{[T_k((e^{-2s});x) - e^{-2x}]$$
$$-2e^{-x}[T_k(e^{-s};x) - e^{-x}] + e^{-2x}[T_k(1;x) - 1]\} + f(x)[T_k(1;x) - 1]$$
$$= \varepsilon[T_k(1;x) - 1] + \varepsilon + \frac{2M}{\delta^2}\{[T_k((e^{-2s});x) - e^{-2x}] - 2e^{-x}[T_k(e^{-s};x) - e^{-x}]$$
$$+ e^{-2x}[T_k(1;x) - 1]\} + f(x)[T_k(1;x) - 1].$$

Since $\varepsilon$ is arbitrary, we can write

$$T_k(f;x) - f(x) \le \varepsilon[T_k(1;x) - 1] + \frac{2M}{\delta^2}\{[T_k(e^{-2s};x) - e^{-2x}]$$
$$-2e^{-x}[T_k(e^{-s};x) - e^{-x}] + e^{-2x}[T_k(1;x) - 1]\} + f(x)[T_k(1;x) - 1].$$

Therefore

$$|T_k(f;x) - f(x)| \le \varepsilon + (\varepsilon + M)|T_k(1;x) - 1| + \frac{2M}{\delta^2}|e^{-2x}||T_k(1;x,y) - 1|$$
$$+\frac{2M}{\delta^2}|T_k(e^{-2s};x)| - e^{-2x}| + \frac{4M}{\delta^2}|e^{-x}||T_k(e^{-s};x) - e^{-x}|$$
$$\le \varepsilon + \left(\varepsilon + M + \frac{4M}{\delta^2}\right)|T_k(1;x) - 1| + \frac{2M}{\delta^2}|e^{-2x}||T_k(1;x) - 1|$$
$$+\frac{2M}{\delta^2}|T_k(e^{-2s};x) - e^{-2x}| + \frac{4M}{\delta^2}|T_k(e^{-s};x) - e^{-x}|$$

(11.2.11)

since $|e^{-x}| \leq 1$ for all $x \in I$. Now, taking $\sup_{x \in I}$, we get

$$\|T_k(f; x) - f(x)\|_\infty$$
$$\leq \varepsilon + K(\|T_k(1; x) - 1\|_\infty + \|T_k(e^{-s}; x) - e^{-x}\|_\infty + \|T_k(e^{-2s}; x) - e^{-2x}\|_\infty,$$

$$(11.2.12)$$

where $K = \max\left\{\varepsilon + M + \frac{4M}{\delta^2}, \frac{2M}{\delta^2}\right\}$. Now replacing $T_k(\cdot, x)$ by $\sum_{k=0}^m T_k(\cdot, x)/(m+1)$ and then by $B_m(\cdot, x)$ in (11.2.12) on both sides. For a given $r > 0$ choose $\varepsilon' > 0$ such that $\varepsilon' < r$. Define the following sets

$$D = \{m \leq n : \|B_m(f, x) - f(x)\|_\infty \geq r\},$$

$$D_1 = \left\{m \leq n : \|B_m(1, x) - 1\|_\infty \geq \frac{r - \varepsilon'}{4K}\right\},$$

$$D_2 = \left\{m \leq n : \|B_m(t, x) - e^{-x}\|_\infty \geq \frac{r - \varepsilon'}{4K}\right\},$$

$$D_3 = \left\{m \leq n : \|B_m(t^2, x) - e^{-2x}\|_\infty \geq \frac{r - \varepsilon'}{4K}\right\}.$$

Then, $D \subset D_1 \cup D_2 \cup D_3$, and so $\delta(D) \leq \delta(D_1) + \delta(D_2) + \delta(D_3)$. Therefore, using conditions (11.2.3)–(11.2.5), we get

$$C_1(\text{st}) - \lim_{n \to \infty} \|T_n(f, x) - f(x)\|_\infty = 0.$$

This completes the proof of the theorem. $\qquad \square$

In the following example we construct a sequence of positive linear operators satisfying the conditions of Theorem 11.2.3 but does not satisfy the conditions of Theorem 11.2.2 as well as its statistical version.

*Example 11.2.4.* Consider the sequence of *classical Baskakov operators[14]*.

$$V_n(f; x) := \sum_{k=0}^\infty f\left(\frac{k}{n}\right)\binom{n - 1 + k}{k} x^k (1 + x)^{-n-k};$$

where $0 \leq x, y < \infty$.
Let $L_n : C(I) \to C(I)$ be defined by

$$L_n(f; x) = |(1 + x_n)V_n(f; x)|,$$

where the sequence $x = (x_n)$ is defined by (11.2.1). Note that this sequence is statistically summable $(C, 1)$ to 0 but neither convergent nor statistically convergent.

Now,

$$L_n(1;x) = 1,$$

$$L_n(e^{-s};x) = (1 + x - xe^{-\frac{1}{n}})^{-n},$$

$$L_n(e^{-2s};x^2) = (1 + x^2 - x^2 e^{-\frac{1}{n}})^{-n},$$

we have that the sequence $(L_n)$ satisfies the conditions (11.2.3)–(11.2.5). Hence by Theorem 11.2.3, we have

$$C_1(\text{st}) - \lim_{n \to \infty} \|L_n(f) - f\|_\infty = 0.$$

On the other hand, we get $L_n(f;0) = |(1 + x_n)f(0)|$, since $V_n(f;0) = f(0)$, and hence

$$\|L_n(f;x) - f(x)\|_\infty \geq |L_n(f;0) - f(0)| = |x_n f(0)|.$$

We see that $(L_n)$ does not satisfy the conditions of the theorem of Boyanov and Veselinov as well as its statistical version, since $(x_n)$ is neither convergent nor statistically convergent. Hence Theorem 11.2.3 is stronger than Theorem 11.2.2 as well as its statistical version.


## 11.3   Application of Statistical $A$-Summability

Let $H_\omega(I)$ denote the space of all real-valued functions $f$ on $I$ such that

$$|f(s) - f(x)| \leq \omega\left(f; \left|\frac{s}{1+s} - \frac{x}{1+x}\right|\right),$$

where $\omega$ is the modulus of continuity, i.e.,

$$\omega(f;\delta) = \sup_{s,x \in I} \{|f(s) - f(x)| : |s - x| \leq \delta\}.$$

It is to be noted that any function $f \in H_\omega(I)$ is continuous and bounded on $I$.

The following Korovkin-type theorem was proved by Çakar and Gadjiev [21].

**Theorem 11.3.1.** *Let $(L_n)$ be a sequence of positive linear operators from $H_\omega(I)$ into $C_B(I)$. Then for all $f \in H_\omega(I)$*

$$\lim_{n \to \infty} \|L_n(f;x) - f(x)\|_{C_B(I)} = 0$$

*if and only if*

$$\lim_{n\to\infty} \|L_n(f_i;x) - g_i\|_{C_B(I)} = 0; \quad (i = 0,1,2),$$

*where*

$$g_0(x) = 1, \ g_1(x) = \frac{x}{1+x}, \ g_2(x) = \left(\frac{x}{1+x}\right)^2.$$

Erkuş and Duman [32] have given the $A$-statistical version of the above theorem for functions of two variables. In this section, we use the notion of statistical $A$-summability to prove a Korovkin-type approximation theorem for functions of two variables with the help of test functions $1, x/(1+x), y/(1+y), [x/(1+x)]^2 + [y/(1+y)]^2$.

Let $I = [0,\infty)$ and $K = I \times I$. We denote by $C_B(K)$ the space of all bounded and continuous real-valued functions on $K$ equipped with norm

$$\|f\|_{C_B(K)} := \sup_{(x,y)\in K} |f(x,y)|, \ f \in C_B(K).$$

Let $H_{\omega^*}(K)$ denote the space of all real-valued functions $f$ on $K$ such that

$$|f(s,t) - f(x,y)| \le \omega^*\left[f; \sqrt{\left(\frac{s}{1+s} - \frac{x}{1+x}\right)^2 + \left(\frac{t}{1+t} - \frac{y}{1+y}\right)^2}\right],$$

where $\omega^*$ is the modulus of continuity, i.e.,

$$\omega^*(f;\delta) = \sup_{(s,t),(x,y)\in K} \{|f(s,t) - f(x,y)| : \sqrt{(s-x)^2 + (t-y)^2} \le \delta\}.$$

It is to be noted that any function $f \in H_{\omega^*}(K)$ is bounded and continuous on $K$, and a necessary and sufficient condition for $f \in H_{\omega^*}(K)$ is that $\omega^*(f;\delta) \to 0$, as $\delta \to 0$.

**Theorem 11.3.2.** *Let $A = (a_{nk})$ be nonnegative regular summability matrix. Let $(T_k)$ be a sequence of positive linear operators from $H_{\omega^*}(K)$ into $C_B(K)$. Then for all $f \in H_{\omega^*}(K)$*

$$st - \lim_{n\to\infty} \left\|\sum_{k=1}^{\infty} a_{nk} T_k(f;x,y) - f(x,y)\right\|_{C_B(K)} = 0 \qquad (11.3.1)$$

*if and only if*

$$st - \lim_{n\to\infty} \left\|\sum_{k=1}^{\infty} a_{nk} T_k(1;x,y) - 1\right\|_{C_B(K)} = 0 \qquad (11.3.2)$$

$$\text{st} - \lim_{n \to \infty} \left\| \sum_{k=1}^{\infty} a_{nk} T_k \left( \frac{s}{1+s}; x, y \right) - \frac{x}{1+x} \right\|_{C_B(K)} = 0, \quad (11.3.3)$$

$$\text{st} - \lim_{n \to \infty} \left\| \sum_{k=1}^{\infty} a_{nk} T_k \left( \frac{t}{1+t}; x, y \right) - \frac{y}{1+y} \right\|_{C_B(K)} = 0, \quad (11.3.4)$$

$$\text{st} - \lim_{n \to \infty} \left\| \sum_{k=1}^{\infty} a_{nk} T_k \left[ \left( \frac{s}{1+s} \right)^2 + \left( \frac{t}{1+t} \right)^2; x, y \right] \right.$$

$$\left. - \left[ \left( \frac{x}{1+x} \right)^2 + \left( \frac{y}{1+y} \right)^2 \right] \right\|_{C_B(K)} = 0. \quad (11.3.5)$$

*Proof.* Since each of the functions $f_0(x, y) = 1$, $f_1(x, y) = x/(1+x)$, $f_2(x, y) = y/(1+y)$, $f_3(x, y) = [x/(1+x)]^2 + [y/(1+y)]^2$ belongs to $H_{\omega*}(K)$, conditions (11.3.2)–(11.3.5) follow immediately from (11.3.1). Let $f \in H_{\omega*}(K)$ and $(x, y) \in K$ be fixed. Then for $\varepsilon > 0$ there exist $\delta_1, \delta_2 > 0$ such that $|f(s, t) - f(x, y)| < \varepsilon$ holds for all $(s, t) \in K$ satisfying $|\frac{s}{1+s} - \frac{x}{1+x}| < \delta_1$ and $|\frac{t}{1+t} - \frac{y}{1+y}| < \delta_2$. Let

$$K(\delta) := \left\{ (s, t) \in K : \sqrt{\left( \frac{s}{1+s} - \frac{x}{1+x} \right)^2 + \left( \frac{t}{1+t} - \frac{y}{1+y} \right)^2} < \delta \right\},$$

where $\delta = \min\{\delta_1, \delta_2\}$. Hence,

$$|f(s, t) - f(x, y)| = |f(s, t) - f(x, y)| \chi_{K(\delta)}(s, t)$$
$$+ |f(s, t) - f(x, y)| \chi_{K \setminus K(\delta)}(s, t)$$
$$\leq \varepsilon + 2N \chi_{K \setminus K(\delta)}(s, t), \quad (11.3.6)$$

where $\chi_D$ denotes the characteristic function of the set $D$ and $N = \|f\|_{C_B(K)}$. Further we get

$$\chi_{K \setminus K(\delta)}(s, t) \leq \frac{1}{\delta_1^2} \left( \frac{s}{1+s} - \frac{x}{1+x} \right)^2 + \frac{1}{\delta_2^2} \left( \frac{t}{1+t} - \frac{y}{1+y} \right)^2. \quad (11.3.7)$$

Combining (11.3.6) and (11.3.7), we get

$$|f(s, t) - f(x, y)| \leq \varepsilon + \frac{2N}{\delta^2} \left[ \left( \frac{s}{1+s} - \frac{x}{1+x} \right)^2 + \left( \frac{t}{1+t} - \frac{y}{1+y} \right)^2 \right].$$

$$(11.3.8)$$

After using the properties of $f$, a simple calculation gives that

$$|T_k(f;x,y)-f(x,y)|\leq\varepsilon+M\{|T_k(f_0;x,y)-f_0(x,y)|+|T_k(f_1;x,y)-f_1(x,y)|$$
$$+|T_k(f_2;x,y)-f_2(x,y)|+|T_k(f_3;x,y)-f_3(x,y)|\}, \qquad (11.3.9)$$

where $M := \varepsilon + N + 4N/\delta^2$. Now replacing $T_k(f;x,y)$ by $\sum_{k=1}^{\infty} a_{nk} T_k(f;x,y)$ and taking $\sup_{(x,y)\in K}$, we get

$$\left\|\sum_{k=1}^{\infty} a_{nk} T_k(f;x,y)-f(x,y)\right\|_{C_B(K)} \leq \varepsilon+M\left[\left\|\sum_{k=1}^{\infty} a_{nk} T_k(f_0;x,y)-f_0(x,y)\right\|_{C_B(K)}\right.$$

$$+\left\|\sum_{k=1}^{\infty} a_{nk} T_k(f_1;x,y)-f_1(x,y)\right\|_{C_B(K)} +\left\|\sum_{k=1}^{\infty} a_{nk} T_k(f_2;x,y)-f_2(x,y)\right\|_{C_B(K)}$$

$$\left.+\left\|\sum_{k=1}^{\infty} a_{nk} T_k(f_3;x,y)-f_3(x,y)\right\|_{C_B(K)}\right]. \qquad (11.3.10)$$

For a given $r > 0$ choose $\varepsilon > 0$ such that $\varepsilon < r$, define the following sets

$$D := \left\{n : \left\|\sum_{k=1}^{\infty} a_{nk} T_k(f;x,y)-f(x,y)\right\|_{C_B(K)} \geq r\right\},$$

$$D_1 := \left\{n : \left\|\sum_{k=1}^{\infty} a_{nk} T_k(f_0;x,y)-f_0(x,y)\right\|_{C_B(K)} \geq \frac{r-\varepsilon}{4K}\right\},$$

$$D_2 := \left\{n : \left\|\sum_{k=1}^{\infty} a_{nk} T_k(f_1;x,y)-f_1(x,y)\right\|_{C_B(K)} \geq \frac{r-\varepsilon}{4K}\right\},$$

$$D_3 := \left\{n : \left\|\sum_{k=1}^{\infty} a_{nk} T_k(f_2;x,y)-f_2(x,y)\right\|_{C_B(K)} \geq \frac{r-\varepsilon}{4K}\right\},$$

$$D_4 := \left\{n : \left\|\sum_{k=1}^{\infty} a_{nk} T_k(f_3;x,y)-f_3(x,y)\right\|_{C_B(K)} \geq \frac{r-\varepsilon}{4K}\right\}.$$

Then from (11.3.10), we see that $D \subset D_1 \cup D_2 \cup D_3 \cup D_4$ and therefore $\delta(D) \leq \delta(D_1) + \delta(D_2) + \delta(D_3) + \delta(D_4)$. Hence the conditions (11.3.2)–(11.3.5) imply the condition (11.3.1).

This completes the proof of the theorem.                                          □

If we replace the matrix $A$ in Theorem 11.3.2 by an identity matrix, then we immediately get the following result which is due to Erkuş and Duman [32]:

**Corollary 11.3.3.** *Let $(T_k)$ be a sequence of positive linear operators from $H_{\omega^*}(K)$ into $C_B(K)$. Then for all $f \in H_{\omega^*}(K)$*

$$\text{st} - \lim_{k \to \infty} \|T_k(f; x, y) - f(x, y)\|_{C_B(K)} = 0 \tag{11.3.11}$$

*if and only if*

$$\text{st} - \lim_{k \to \infty} \|T_k(1; x, y) - 1\|_{C_B(K)} = 0, \tag{11.3.12}$$

$$\text{st} - \lim_{k \to \infty} \left\| T_k\left(\frac{s}{1+s}; x, y\right) - \frac{x}{1+x} \right\|_{C_B(K)} = 0, \tag{11.3.13}$$

$$\text{st} - \lim_{k \to \infty} \left\| T_k\left(\frac{t}{1+t}; x, y\right) - \frac{y}{1+y} \right\|_{C_B(K)} = 0, \tag{11.3.14}$$

$$\text{st} - \lim_{k \to \infty} \left\| T_k\left[\left(\frac{s}{1+s}\right)^2 + \left(\frac{t}{1+t}\right)^2; x, y\right] \right.$$

$$\left. - \left[\left(\frac{x}{1+x}\right)^2 + \left(\frac{y}{1+y}\right)^2\right] \right\|_{C_B(K)} = 0. \tag{11.3.15}$$

*Example 11.3.4.* We show that the following double sequence of positive linear operators satisfies the conditions of Theorem 11.3.2 but does not satisfy the conditions of Corollary 11.3.3 and Theorem 11.3.1.

Consider the following Bleimann, Butzer, and Hahn [16] (of two variables) operators:

$$B_n(f; x, y)$$

$$:= \frac{1}{(1+x)^n (1+y)^n} \sum_{j=0}^{n} \sum_{k=0}^{n} f\left(\frac{j}{n-j+1}, \frac{k}{n-k+1}\right) \binom{n}{j}\binom{n}{k} x^j y^k, \tag{11.3.16}$$

where $f \in H_\omega(K)$, $K = [0, \infty) \times [0, \infty)$ and $n \in \mathbb{N}$. Since

$$(1+x)^n = \sum_{j=0}^{n} \binom{m}{j} x^j,$$

it is easy to see that

$$\lim_{n \to \infty} B_n(f_0; x, y) = 1 = f_0(x, y).$$

Also by a simple calculation, we obtain

$$\lim_{n \to \infty} B_n(f_1; x, y) = \lim_{n \to \infty} \frac{n}{n+1} \left( \frac{x}{1+x} \right) = \frac{x}{1+x} = f_1(x, y),$$

$$\lim_{n \to \infty} B_n(f_2; x, y) = \lim_{n \to \infty} \frac{n}{n+1} \left( \frac{y}{1+y} \right) = \frac{y}{1+y} = f_2(x, y).$$

Finally, we get

$$\lim_{n \to \infty} B_n(f_3; x, y) = \lim_{n \to \infty} \left[ \frac{n(n-1)}{(n+1)^2} \left( \frac{x}{1+x} \right)^2 + \frac{n}{(n+1)^2} \left( \frac{x}{1+x} \right) \right.$$

$$\left. + \frac{n(n-1)}{(n+1)^2} \left( \frac{y}{1+y} \right)^2 + \frac{n}{(n+1)^2} \left( \frac{y}{1+y} \right) \right]$$

$$= \left( \frac{x}{1+x} \right)^2 + \left( \frac{y}{1+y} \right)^2 = f_3(x, y).$$

Now, take $A = (C, 1)$ and define $u = (u_n)$ by

$$u_k = \begin{cases} 1, & k \text{ is odd}, \\ 0, & k \text{ is even}. \end{cases}$$

Let the operator $L_n : H_\omega(K) \to C_B(K)$ be defined by

$$L_n(f; x, y) = (1 + u_n) B_n(f; x, y).$$

Then the sequence $(L_n)$ satisfies the conditions (11.3.2)–(11.3.5). Hence by Theorem 11.3.2, we have

$$\text{st} - \lim_{m \to \infty} \left\| \sum_{n=1}^{\infty} a_{mn} L_n(f; x, y) - f(x, y) \right\|_{C_B(K)}$$

$$= \text{st} - \lim_{m \to \infty} \left\| \frac{1}{m} \sum_{n=1}^{m} L_n(f; x, y) - f(x, y) \right\|_{C_B(K)} = 0.$$

On the other hand, the sequence $(L_n)$ does not satisfy the conditions of Theorem 11.3.1, Corollary 11.3.3, and Theorem 2.1 of [32], since $(L_n)$ is neither convergent nor statistically (nor $A$-statistically) convergent. That is, Theorem 11.3.1, Corollary 11.3.3, and Theorem 2.1 of [32] do not work for our operators $L_n$. Hence Theorem 11.3.2 is stronger than Corollary 11.3.3 and Theorem 2.1 of [32].

## 11.4   Rate of Statistical $A$-Summability

In this section, using the concept of statistical $A$-summability, we study the rate of convergence of positive linear operators with the help of the modulus of continuity. Let us recall, for $f \in H_{\omega^*}(K)$

$$|f(s,t) - f(x,y)| \leq \omega^* \left[ f; \sqrt{\left(\frac{s}{1+s} - \frac{x}{1+x}\right)^2 + \left(\frac{t}{1+t} - \frac{y}{1+y}\right)^2} \right],$$

where

$$\omega^*(f;\delta) = \sup_{(s,t),(x,y)\in K} \{|f(s,t) - f(x,y)| : \sqrt{(s-x)^2 + (t-y)^2} \leq \delta\}.$$

We have the following result:

**Theorem 11.4.1.** *Let $A = (a_{nk})$ be nonnegative regular summability matrix and $(T_k)$ be a sequence of positive linear operators from $H_{\omega^*}(K)$ into $C_B(K)$. Assume that*

*(i)* $\mathrm{st} - \lim_{n\to\infty} \left\| \sum_{k=1}^{\infty} a_{nk} T_k(f_0) - f_0 \right\|_{C_B(K)} = 0,$
*(ii)* $\mathrm{st} - \lim_{n\to 0} \omega^*(f;\delta_n) = 0,$

*where*

$$\delta_n = \sqrt{\left\| \sum_{k=1}^{\infty} a_{nk} T_k(\psi) \right\|_{C_B(K)}} \quad \text{with } \psi = \psi(s,t) = \left(\frac{s}{1+s} - \frac{x}{1+x}\right)^2 + \left(\frac{t}{1+t} - \frac{y}{1+y}\right)^2.$$

*Then for all $f \in H_{\omega^*}(K)$*

$$\mathrm{st} - \lim_{n\to\infty} \left\| \sum_{k=1}^{\infty} a_{nk} T_k(f) - f \right\|_{C_B(K)} = 0.$$

*Proof.* Let $f \in H_{\omega^*}(K)$ be fixed and $(x,y) \in K$ be fixed. Using linearity and positivity of the operators $T_k$ for all $n \in \mathbb{N}$, we have

$$\left| \sum_{k=1}^{\infty} a_{nk} T_k(f;x,y) - f(x,y) \right| \leq \sum_{k=1}^{\infty} a_{nk} T_k(|f(s,t) - f(x,y)|; x,y)$$

$$+ |f(x,y)| \left| \sum_{k=1}^{\infty} a_{nk} T_k(f_0; x,y) - f_0(x,y) \right|$$

$$\leq \sum_{k=1}^{\infty} a_{nk} T_k \left[ \omega^* \left( f; \delta \frac{\sqrt{\left(\frac{s}{1+s} - \frac{x}{1+x}\right)^2 + \left(\frac{t}{1+t} - \frac{y}{1+y}\right)^2}}{\delta} \right); x, y \right]$$

$$+ \|f\|_{C_B(K)} \left| \sum_{k=1}^{\infty} a_{nk} T_k (f_0; x, y) - f_0(x, y) \right|$$

$$\leq \sum_{k=1}^{\infty} a_{nk} T_k \left[ \left( 1 + \frac{\sqrt{\left(\frac{s}{1+s} - \frac{x}{1+x}\right)^2 + \left(\frac{t}{1+t} - \frac{y}{1+y}\right)^2}}{\delta} \right) \omega^*(f; \delta); x, y \right]$$

$$+ \|f\|_{C_B(K)} \left| \sum_{k=1}^{\infty} a_{nk} T_k (f_0; x, y) - f_0(x, y) \right|$$

$$\leq \sum_{k=1}^{\infty} a_{nk} \omega^*(f; \delta) T_k \left[ \left( 1 + \frac{\left(\frac{s}{1+s} - \frac{x}{1+x}\right)^2 + \left(\frac{t}{1+t} - \frac{y}{1+y}\right)^2}{\delta^2}; x, y \right) \right]$$

$$+ \|f\|_{C_B(K)} \left| \sum_{k=1}^{\infty} a_{nk} T_k (f_0; x, y) - f_0(x, y) \right|$$

$$\leq \omega^*(f; \delta) \left| \sum_{k=1}^{\infty} a_{nk} T_k (f_0; x, y) - f_0(x, y) \right| + \|f\|_{C_B(K)} \left| \sum_{k=1}^{\infty} a_{nk} T_k (f_0; x, y) - f_0(x, y) \right|$$

$$+ \omega^*(f; \delta) + \frac{\omega^*(f; \delta)}{\delta^2} \sum_{k=1}^{\infty} a_{nk} T_k \left[ \left( \frac{s}{1+s} - \frac{x}{1+x} \right)^2 + \left( \frac{t}{1+t} - \frac{y}{1+y} \right)^2; x, y \right].$$

Hence,

$$\left\| \sum_{k=1}^{\infty} a_{nk} T_k (f) - f \right\|_{C_B(K)}$$

$$\leq \|f\|_{C_B(K)} \left\| \sum_{k=1}^{\infty} a_{nk} T_k (f_0) - f_0 \right\|_{C_B(K)} + \omega^*(f; \delta) \left\| \sum_{k=1}^{\infty} a_{nk} T_k (f_0) - f_0 \right\|_{C_B(K)}$$

$$+ \frac{\omega^*(f; \delta)}{\delta^2} \left\| \sum_{k=1}^{\infty} a_{nk} T_k (\psi) \right\|_{C_B(K)} + \omega^*(f; \delta).$$

Now if we choose $\delta := \delta_n := \sqrt{\left\| \sum_{k=1}^{\infty} a_{nk} T_k (\psi) \right\|_{C_B(K)}}$, then

$$\left\| \sum_{k=1}^{\infty} a_{nk} T_k(f) - f \right\|_{C_B(K)}$$

$$\leq \|f\|_{C_B(K)} \left\| \sum_{k=1}^{\infty} a_{nk} T_k(f_0) - f_0 \right\|_{C_B(K)}$$

$$+ \omega^*(f;\delta_n) \left\| \sum_{k=1}^{\infty} a_{nk} T_k(f_0) - f_0 \right\|_{C_B(K)} + 2\omega^*(f;\delta_n).$$

Therefore,

$$\left\| \sum_{k=1}^{\infty} a_{nk} T_k(f) - f \right\|_{C_B(K)} \leq M \left\{ \left\| \sum_{k=1}^{\infty} a_{nk} T_k(f_0) - f_0 \right\|_{C_B(K)} \right.$$

$$\left. + \omega^*(f;\delta_n) \left\| \sum_{k=1}^{\infty} a_{nk} T_k(f_0) - f_0 \right\|_{C_B(K)} + \omega^*(f;\delta_n) \right\}, \qquad (11.4.1)$$

where $M = \max\{2, \|f\|_{C_B(K)}\}$. Now, for a given $r > 0$, choose $\varepsilon > 0$ such that $\varepsilon > r$. Let us write

$$E := \left\{ n : \left\| \sum_{k=1}^{\infty} a_{nk} T_k(f;x,y) - f(x,y) \right\|_{C_B(K)} \geq r \right\},$$

$$E_1 := \left\{ n : \left\| \sum_{k=1}^{\infty} a_{nk} T_k(f_0;x,y) - f_0(x,y) \right\|_{C_B(K)} \geq \frac{r}{3K} \right\},$$

$$E_2 := \left\{ n : \omega^*(f;\delta_n) \geq \frac{r}{3K} \right\},$$

$$E_3 := \left\{ n : \omega^*(f;\delta_n) \left\| \sum_{k=1}^{\infty} a_{nk} T_k(f_0;x,y) - f_0(x,y) \right\|_{C_B(K)} \geq \frac{r}{3K} \right\}.$$

Then $E \subset E_1 \cup E_2 \cup E_3$ and therefore $\delta(E) \leq \delta(E_1) + \delta(E_2) + \delta(E_3)$. Using conditions (i) and (ii) we conclude

$$\mathrm{st} - \lim_{n \to \infty} \left\| \sum_{k=1}^{\infty} a_{nk} T_k(f) - f \right\|_{C_B(K)} = 0.$$

This completes the proof of the theorem.                                        $\square$

# Chapter 12
# Applications to Fixed Point Theorems

## 12.1 Introduction

Let $E$ be a closed, bounded, convex subset of a Banach space $X$ and $f : E \longrightarrow E$. Consider the iteration scheme defined by $\bar{x}_0 = x_0 \in E$, $\bar{x}_{n+1} = f(x_n)$, $x_n = \sum_{k=0}^{n} a_{nk}\bar{x}_k$, $n \geq 1$, where $A$ is a regular weighted mean matrix. For particular spaces $X$ and functions $f$ we show that this iterative scheme converges to a fixed point of $f$. During the past few years several mathematicians have obtained fixed point results using Mann and other iteration schemes for certain classes of infinite matrices. In this chapter, we present some results using such schemes which are represented as regular weighted mean methods. Results of this chapter appeared in [20, 40, 82] and [84].

## 12.2 Definitions and Notations

Let $E$ be a nonempty closed convex subset of a Banach space $X$. A mapping $T : E \to E$ is said to be

(a) a *contraction* on $X$, if there is some nonnegative real number $k < 1$ such that for all $x$ and $y$ in $E$, $\|Tx - Ty\| \leq k\|x - y\|$;
(b) a *non-expansive map* if $\|Tx - Ty\| \leq \|x - y\|$;
(c) *quasi non-expansive map* if

$\|Tx - Ty\|$

$$\leq a_1\|x - y\| + a_2\|x - Tx\| + a_3\|y - Ty\| + a_4\|x - Ty\| + a_5\|y - Tx\| \quad (*)$$

for all $x, y \in E$, $a_i \geq 0$ and $\sum_{i=1}^{5} a_i \leq 1$.

M. Mursaleen, *Applied Summability Methods*, SpringerBriefs in Mathematics,
DOI 10.1007/978-3-319-04609-9_12, © M. Mursaleen 2014

Let $(X, d)$ be a metric space and $T : X \to X$ be a mapping. The point $x \in X$ is called a *fixed point* of $T$ if $Tx = x$.

The following generalized iteration process is called *Mann iteration* in which $A = (a_{ij})$, $i, j \in \mathbb{N}$ is an infinite matrix of real numbers such that

$$a_{ij} \geq 0, \text{ for all } i, j \in \mathbb{N} \text{ and } a_{ij} = 0 \text{ for } j > i, \qquad (12.2.1)$$

$$\lim_{i \to \infty} a_{ij} = 0 \text{ for each fixed } j \in \mathbb{N}, \qquad (12.2.2)$$

$$\sum_{j=1}^{i} a_{ij} = 1, \text{ for all } i \in \mathbb{N}. \qquad (12.2.3)$$

Obviously the above matrix $A$ is regular. If $E$ be a nonempty, closed, convex subset of a Banach space $B$ and $T$ be a mapping of $E$ into itself satisfying certain conditions, then starting with an arbitrary element $x_1 \in E$, the generalized iteration process, denoted by the triplet $(x_1, A, T)$, is defined by

$$x_{n+1} = Tv_n \text{ where } v_n = \sum_{k=1}^{n} a_{nk} x_k, \text{ for all } n \in \mathbb{N}.$$

Various choice of the infinite matrix $A$ yields many interesting iterative process as special cases. Taking $A$ to be the infinite identity matrix $I$, then the process $(x_1, A, T)$ is just an ordinary Picard iteration defined by

$$v_{n+1} = x_{n+1} = Tv_n \text{ whence } v_{n+1} = T^n v_1 = T^n x_1.$$

In many particular problems the generalized iteration process can easily be seen to converge while the ordinary Picard iteration process may not converge.

Let $X$ be a Banach space. A sequence $(x_n)$ in $X$ is said to be (a) *almost (strongly) convergent* to $z \in X$ if the strong $\lim_{n \to \infty} \frac{1}{n} \sum_{j=k}^{k+n-1} x_j = z$ uniformly in $k$, (b) *almost weakly convergent* to $z \in X$ if $\langle x_n, y \rangle$ is almost convergent to $\langle x, y \rangle$ for all $y \in X^*$.

Let $E$ be a nonempty closed and convex subset of a Banach space $X$ and $\{x_n\}$ a bounded sequence in $X$. For $x \in X$, define the *asymptotic radius* of $\{x_n\}$ at $x$ as the number $r(x, \{x_n\}) = \limsup_{n \to \infty} \| x_n - x \|$. Let $r = r(E, \{x_n\}) := \inf\{r(x, \{x_n\}) : x \in E\}$ and $A = A(E, \{x_n\}) := \{x \in E : r(x, \{x_n\}) = r\}$. The number $r$ and the set $A$ are called the *asymptotic radius* and *asymptotic center* relative to $E$, respectively.

## 12.3   Iterations of Regular Matrices

Let $X$ be a normed linear space, $E$ a nonempty closed bounded, convex subset of $X$, $f : E \longrightarrow E$ possessing at least one fixed point in $E$, and $A$ an infinite matrix. Given the iteration scheme

$$\bar{x}_0 = x_0 \in E, \tag{12.3.1}$$

$$\bar{x}_{n+1} = f(x_n), \quad n = 0, 1, 2, \cdots, \tag{12.3.2}$$

$$x_n = \sum_{k=0}^{n} a_{nk} \bar{x}_k, \quad n = 1, 2, 3, \cdots, \tag{12.3.3}$$

it is reasonable to ask what restrictions on the matrix $A$ are necessary and/or sufficient to guarantee that the above iteration scheme converges to a fixed point of $f$.

Several mathematicians have obtained result using iteration schemes of the form (12.3.1)–(12.3.3) for certain classes of infinite matrices. We shall confine our attention to regular triangular matrices $A$ satisfying:

$$0 \le a_{nk} \le 1, \quad n, k = 0, 1, 2, \cdots, \tag{12.3.4}$$

$$\sum_{k=0}^{n} a_{nk} = 1, \quad n = 0, 1, 2, \cdots. \tag{12.3.5}$$

Conditions (12.3.4) and (12.3.5) are obviously necessary in order to ensure that $x_n$ and $\bar{x}_n$ in (12.3.2) and (12.3.3) remain in $E$. The scheme (12.3.1)–(12.3.3) is generally referred to as the Mann process.

Barone [12] observed that a sufficient condition for a regular matrix $A$ to transform each bounded sequence into a sequence whose set of limit points is connected is that $A$ satisfies

$$\lim_{n} \sum_{k=0}^{\infty} |a_{nk} - a_{n-1,k}| = 0. \tag{12.3.6}$$

Rhoades announced the following conjecture:
*Conjecture.* Let $f$ be a continuous mapping of $[a, b]$ into itself, $A$ a regular matrix satisfying (12.3.4)–(12.3.6). Then the iteration scheme defined by (12.3.1)–(12.3.3) converges to a fixed point of $f$.

The conjecture need not remain true if condition (12.3.6) is removed. To see this, let $A$ be the identity matrix, $[a, b] = [0, 1]$, $f(x) = 1 - x$, and choose $x_0 = 0$.

The conjecture is true for a large class of weighted mean matrices as we now show.

A weighted mean method is a regular triangular method $A = (a_{nk})$ defined by $a_{nk} = p_k / P_n$, where the sequence $\{p_n\}$ satisfies $p_0 > 0$, $p_n \ge 0$ for $n > 0$, $P_n = \sum_{k=0}^{n} p_k$, and $P_n \to \infty$ as $n \to \infty$. It is easy to verify [84] that such a matrix satisfies (12.3.6) if and only if $p_n / P_n \to 0$ as $n \to \infty$.

**Theorem 12.3.1.** *Let $A$ be a regular weighted mean method satisfying (12.3.6), $f$ a continuous mapping from $E = [a, b]$ into itself. Then the iteration scheme (12.3.1)–(12.3.3) converges to a fixed point of $f$.*

*Proof.* There is no loss of generality in assuming $[a, b] = [0, 1]$. Any regular weighted mean method automatically satisfies conditions (12.3.4) and (12.3.5). Using (12.3.3) we may write

$$x_{n+1} = (p_{n+1}/P_{n+1})(f(x_n) - x_n) + x_n. \tag{12.3.7}$$

Since $x_n, f(x_n) \in [0, 1]$, we have, from (12.3.7), $|x_{n+1} - x_n| \le p_{n+1}/P_{n+1} \to 0$ as $n \to \infty$.

Now following the proof in ([35], p.325), we can easily establish that $\{x_n\}$ converges. It remains to show that $\{x_n\}$ tends to a fixed point of $f$.

*Fact.* Let $A$ be any regular matrix, $f$ as defined above. If the iteration scheme (12.3.1)–(12.3.3) converges, it converges to a fixed point of $f$.

Let $x = \{x_n\}$, $\bar{x} = \{\bar{x}_n\}$, $y = \lim_n \bar{x}_n = f(y)$. But $A$ is a regular matrix. Hence $y = \lim_n x_n = \lim_n A_n(\bar{x}) = f(y)$. $\qquad\qquad\qquad\qquad\square$

**Remark 12.3.2.** One obtains the theorem of [35] by setting $p_n = 1$ in Theorem 12.3.1.

Reiermann [83] defines a summability matrix $A = (a_{nk})$ by

$$a_{nk} = \begin{cases} c_k \displaystyle\prod_{j=k+1}^{n} (1 - c_j) \,, & k < n, \\[2mm] c_n & , \ k = n, \\[2mm] 0 & , \ k > n, \end{cases} \tag{12.3.8}$$

where the real sequence $\{c_n\}$ satisfies (i) $c_0 = 1$, (ii) $0 < c_n < 1$ for $n \ge 1$, and (iii) $\sum_k c_k$ diverges. (it is easy to verify [84] that $A$ is regular and satisfies conditions (12.3.4) and (12.3.5). Actually Reinermann permits $c_n = 1$ in order to take care of the identity matrix, but in all interesting applications the restriction $c_n < 1$ is imposed). He then defines the iteration scheme (12.3.1) and $x_{n+1} = \sum_{k=0}^{n} a_{nk} f(x_k)$, which can be written in the form

$$x_{n+1} = (1 - c_n)x_n + c_n f(x_n), \tag{12.3.9}$$

and establishes the following.

**Theorem 12.3.3 ([83], p.211).** *Let $a, b \in R$, $a < b$, $E = [a, b]$, $f : E \to E$, $f$ continuous and with at most one fixed point. With $A$ as defined in (12.3.8) and with $\{c_n\}$ satisfying (i)–(iii) and $\lim_n c_n = 0$, the iteration scheme (12.3.1), (12.3.9) converges to the fixed point of $f$.*

**Theorem 12.3.4.** *The matrix of (12.3.8) with $\{c_n\}$ satisfying (i)–(iii) is a regular weighted mean matrix.*

*Proof.* For, set $a_{nk} = p_k/P_n$, $k \le n$. Then $p_k/p_{k+1} = a_{nk}/a_{n,k+1} = c_k(1 - c_{k+1})/c_{k+1}$, which can be solved to obtain

$$p_k = c_k p_0/ \prod_{j=1}^{k}(1 - c_j), \quad k > 0. \tag{12.3.10}$$

By induction one can show that $P_n = p_0/ \prod_{j=1}^{n}(1 - c_j)$, $n > 0$. Since $\sum_k c_k$ diverges, the product must diverge to 0. Therefore $P_n \to \infty$ as $n \to \infty$ and the weighted mean method $(\bar{N}, p_n)$ with $p_n$ defined by (12.3.10) is regular. Also, each $p_k > 0$.

Conversely, let $(\bar{N}, p_n)$ be a regular weighted mean method with each $p_k > 0$ and define $\{c_n\}$ by

$$c_n = p_n/P_n, \quad n \ge 0. \tag{12.3.11}$$

Then $c_0 = 1$, and, since each $p_k > 0$, $0 < c_n < 1$ for all $n > 0$. Now from (12.3.11), $1 - c_n = P_{n-1}/P_n$, which leads to $P_n = p_0/ \prod_{j=1}^{n}(1 - c_j)$. Therefore $p_k/P_n = c_k \prod_{j=k+1}^{n}(1 - c_j)$ and $A$ has the form (12.3.8). Moreover, $\sum_k c_k$ diverges because $P_n \to \infty$ as $n \to \infty$. Since $c_n = p_n/P_n$, the condition $\lim_n c_n = 0$ is the same as $(\bar{N}, p_n)$ satisfying (12.3.6). $\qquad\square$

*Remark 12.3.5.* We point out, however, that even though matrices involved are the same, the iteration schemes (12.3.1)–(12.3.3) and (12.3.1), (12.3.9) are different. Scheme (12.3.1)–(12.3.3) takes the form $x = Az$, where $z = \{x_0, f(x_0), f(x_1), \cdots\}$; whereas (12.3.1) and (12.3.9) become $x = Aw$, where $w = \{f(x_0), f(x_1), \cdots\}$. In other words the first scheme uses a translate of $w$. However, since $f$ is continuous, it is easy [84] to verify, using the fact, that each method converges to the same fixed point.

Hillam [45] has shown the conjecture to be false and has established the following result, which is a slight generalization of Theorem 12.3.1.

**Theorem 12.3.6 ([45], p.16).** *Let* $f : [0, 1] \to [0, 1]$, *f continuous, A a regular triangular matrix satisfying (12.3.4)–(12.3.6) and*

$$\sum_{k=0}^{n} |a_{n+1,k} - (1 - a_{n+1,k+1})a_{nk}| = o(a_{n+1,n+1}).$$

*If, in addition,*

$$\sum_{n=1}^{\infty}\sum_{k=0}^{n}|a_{n+1,k} - (1 - a_{n+1,k+1})a_{nk}| < \infty,$$

*then the iteration scheme (12.3.1)–(12.3.3) converges to a fixed point of $f$.*

**Theorem 12.3.7.** *Let $E$ be nonempty closed convex subset of a Banach space $B$ and let $T : E \to E$ be a mapping satisfying condition $(*)$. If for any $x_1 \in E$ and a generalized iteration process $(x_1, A, T)$ such that the sequences $\{x_n\}_n$ and $\{v_n\}_n$ both converge to $p$, then $p$ is the unique fixed point of $T$ in $E$.*

*Proof.* Let $x_1 \in E$ and $A$ to be an infinite matrix defined by Mann. In view of (12.3.2), $v_n \in E, \forall n \in \mathbb{N}$ which is assured by the restriction (iii) on $A$.

We now claim that $p = Tp$. If possible, suppose that $p \neq Tp$. Then

$$
\begin{aligned}
\|p - Tp\| &\leq \|x_{n+1} - p\| + \|x_{n+1} - Tp\| \\
&= \|x_{n+1} - p\| + \|Tv_n - Tp\| \\
&\leq \|x_{n+1} - p\| + a_1\|v_n - p\| + a_2\|v_n - Tv_n\| \\
&\quad + a_3\|p - Tp\| + a_4\|v_n - Tp\| + a_5\|p - Tv_n\|.
\end{aligned}
$$

Now passing through the limit as $n \to \infty$, we have
$(1 - a_3 - a_4)\|p - Tp\| \leq 0$.
Interchanging the roles of $Tp$ and $x_{n+1}$, we can have similarly,
$(1 - a_2 - a_5)\|p - Tp\| \leq 0$.
Adding those two, we have
$\{2 - (a_2 + a_3 + a_4 + a_5)\}\|p - Tp\| \leq 0$
which is a contradiction. Hence we must have $p = Tp$.

We shall now show the uniqueness of the fixed point $p$ of $T$. Let $u(\neq p) \in E$ be another fixed point of $T$ in $E$. Thus we have

$$
\begin{aligned}
\|u - p\| &= \|Tu - Tp\| \\
&\leq a_1\|u - p\| + a_2\|u - Tu\| + a_3\|p - Tp\| + a_4\|u - Tp\| + a_5\|p - Tu\| \\
&= (a_1 + a_4 + a_5)\|u - p\| \\
&\leq (1 - a_2 - a_3)\|u - p\|
\end{aligned}
$$

$\Rightarrow (a_2 + a_3)\|u - p\| \leq 0$ which is a contradiction. Therefore $u = p$.
This completes the proof of the theorem.      $\square$

## 12.4  Nonlinear Ergodic Theorems

Brézis and Browder [20] extended Baillon's theorems [9, 10] from the usual Cesàro means of ergodic theory to general averaging processes $A_n = \sum_{k=0}^{\infty} a_{nk} T^k$, where $(a_{nk})_{n,k=0}^{\infty}$ is an infinite matix such that $a_{nk} \geq 0$ and $\sum_{k=0}^{\infty} a_{nk} = 1$. We present here a slight modification as follows.

**Theorem 12.4.1.** *Let $H$ be a Hilbert space, $C$ a closed bounded convex subset of $H$, and $T$ a non-expansive self map of $C$. Let $A = (a_{nk})_{n,k=0}^{\infty}$ be a strongly regular matrix of nonnegative real numbers. Then for each $x \in C$, $A_n x = \sum_{k=0}^{\infty} a_{nk} T^k x$ converges weakly to a fixed point of $T$.*

*Proof.* The following extension of Opial's lemma [79] will be needed in proving this result.

**Lemma 12.4.2.** *Let $\{x_k\}$ and $\{y_k\}$ be two sequences in $H$, $F$ a nonempty subset of $H$, $C_m$ the convex closure of $\cup_{j \geq m}\{x_j\}$. Suppose that*

*(a) for each $f \in F$, $|x_j - f|^2 \to p(f) < +\infty$;*
*(b) $dist(y_k, C_m) \to 0$ as $k \to \infty$ for each $m$;*
*(c) any weak limit of an infinite subsequence of $\{y_k\}$ lies in $F$.*

*Then $y_k$ converges weakly to a point of $F$.*

We apply Lemma 12.4.2 with $F$ the fixed point set of $T$ in $C$, $x_k = T^k x$, $y_n = \sum_{k=0}^{\infty} a_{nk} x_k$. Since $|x_j - f|^2$ decreases with $j$, it converges to $p(f) < +\infty$. Since, by regularity of $A$, $a_{nk} \to 0$ as $n \to +\infty$, $dist(y_n, C_m) \to 0$ as $n \to \infty$ for each $m$. To show that (c) holds, it suffices to prove that $|y_n - T y_n| \to 0$ as $n \to +\infty$. For any $u$ in $H$,

$$| y_n - u |^2 = \left| \sum_{k=0}^{\infty} a_{nk}(x_k - u) \right|^2 = \sum_{j,k=0}^{\infty} a_{nj} a_{nk} \langle x_j - u, x_k - u \rangle.$$

Since

$$2\langle x_j - u, x_k - u \rangle = | x_j - u |^2 + | x_k - u |^2 - | x_j - x_k |^2,$$

$$2 | y_n - u |^2 = 2 \sum_{k=0}^{\infty} a_{nk} | x_k - u |^2 - r_n,$$

where $r_n = \sum_{j,k=0}^{\infty} a_{nj} a_{nk} | x_j - x_k |^2$. If we choose $u = y_n$, then $r_n = \sum_{k=0}^{\infty} a_{nk} | x_k - y_n |^2$. If we take $u = T y_n$, then

$$2 | y_n - T y_n |^2 = 2 a_{n,0} |x - T y_n|^2 + 2 \sum_{k=0}^{\infty} a_{nk} | T x_{k-1} - T y_n |^2 - r_n$$

$$\leq 2 a_{n,0} |x - T y_n|^2 + 2 \sum_{k=0}^{\infty} a_{nk} | x_{k-1} - y_n |^2 - 2 \sum_{k=0}^{\infty} a_{nk} | x_k - y_n |^2$$

$$\leq 2a_{n,0} |x - Ty_n|^2 + 2 \sum_{k=0}^{\infty} (a_{n,k+1} - a_{nk}) \, | \, x_k - y_n \, |^2$$

$$\leq 2 \left( a_{n,0} + \sum_{k=0}^{\infty} | \, a_{n,k+1} - a_{nk} \, | \right) diam(C)^2 \rightarrow 0 \ (n \rightarrow \infty),$$

by strong regularity of $A$.

Hence we get the desired result.                                                              □

Next result is due to Reich [82] in which the notion of almost convergence is used.

**Theorem 12.4.3.** *Let $H$ be a Hilbert space, $C$ a closed bounded convex subset of $H$, and $T$ a non-expansive self map of $C$ with a fixed point. Let $A = (a_{nk})_{n,k=0}^{\infty}$ be a strongly regular matrix of nonnegative real numbers. Then for each $x \in C$, $\{y_n\} = \{A_n x\}$ converges weakly to a fixed point $z$ of $T$ that is the asymptotic center of $\{T^n x\}$.*

*Proof.* Let us write $S_n(x_k) = \frac{1}{n} \sum_{j=k}^{k+n-1} x_j$ for any sequence $\{x_n\}$. Let $F$ be the fixed point set of $T$ and $P : C \rightarrow F$ the nearest point projection. Writing $x_n$ for $T^n x$. Let $\{k(n)\}$ be an arbitrary sequence of natural numbers and $f$ any point in $F$. Note that $\{x_n\}$ is bounded, $\{Px_n\}$ converges strongly to $z$, and

$$\langle S_n(Px_{k(n)} - S_n(x_{k(n)}), f - z \rangle \geq -M S_n(| \, Px_{k(n)} - z \, |)$$

for some constant $M$. Also

$$| \, S_n(x_{k(n)}) - T S_n(x_{k(n)}) \, | \leq \frac{1}{n^{1/2}} \, | \, x_{k(n)} - T S_n(x_{k(n)}) \, | \, .$$

Therefore if $\{S_n(x_{k(n)})\}$ converges weakly to $q$, then we have (i) $\langle z - q, f - z \rangle \geq 0$ for all $f \in F$ and hence $Pq = z$, (ii) $q \in F$. In other words, $S_n(x_{k(n)}) \rightarrow z$ (weakly) and $\{x_n\}$ is almost weakly convergent to $z$. Now applying Theorem 7 of Lorentz [58], strong regularity of $A$ yields the desired result.

This completes the proof of the theorem.                                                     □

# Bibliography

1. F. Altomare, Korovkin-type theorems and approximation by positive linear operators. Surv. Approx. Theor. **5**, 92–164 (2010)
2. R.P. Agnew, Euler transformations. Amer. J. Math. **66**, 318–338 (1944)
3. R.P. Agnew, A simple sufficient condition that a method of summability be stronger than convergence, Bull. Amer. Math. Soc. **52**, 122–132 (1946)
4. M.A. Alghamdi, M. Mursaleen, Hankel matrix transformation of the Walsh-Fourier series, Appl. Math. Comput. **224**, 278–282 (2013)
5. M. Amram, Analytic continuation by summation methods, Isreal J. Maths. **1**, 224–228 (1963)
6. K. Ananda-Rau, On Lambert's series. Proc. London Math. Soc. 2(19), 1–20 (1921)
7. G.A. Anastassiou, M. Mursaleen, S.A. Mohiuddine, Some approximation theorems for functions of two variables through almost convergence of double sequences. J. Comput. Anal. Appl. **13**, 37–40 (2011)
8. B. Bajsanski, Sur une classe generale de procedes de sommations du type D'Euler-Borel. Publ. Inst. Math. (Beograd) **10**, 131–152 (1956)
9. J.-B. Baillon, Un theoreme de type erodique pour les contractions nonlineaires dans un espace de Hilbert. C. R. Acad. Sci. Paris **280**, 1511–1514 (1975)
10. J.-B. Baillon, Quelques proprietes de convergence asymptotique pour les contractions impaire. C. R. Acad. Sci. Paris **283**, 587–590 (1976)
11. S. Banach, *Théorie des Operations Lineaires* (Warszawa, 1932)
12. H.G. Barone, Limit points of sequences and their transforms by methods of summability. Duke Math. J. **5**, 740–752 (1939)
13. F. Başar, *Summability Theory and Its Applications* (Bentham Science Publishers, e-books, Monographs, Istanbul, 2011)
14. M. Becker, Global approximation theorems for Szasz–Mirakjan and Baskakov operators in polynomial weight spaces. Indiana Univ. Math. J. **27**(1), 127–142 (1978)
15. C. Belen, M. Mursaleen and M. Yildirim, Statistical $A$-summability of double sequences and a Korovkin type approximation theorem. Bull. Korean Math. Soc. **49**(4), 851–861 (2012)
16. G. Bleimann, P.L. Butzer, L. Hahn, A Bernstein type operator approximating continuous functions on semiaxis. Indag. Math. **42**, 255–262 (1980)
17. J. Boos, *Classical and Modern Methods in Summability* (Oxford University Press, New York, 2000)
18. E. Borel, *Lecons Sur Iess-Series Divergentes*, 2nd edn. (Horwood, Paris, 1928)
19. B.D. Boyanov, V.M. Veselinov, A note on the approximation of functions in an infinite interval by linear positive operators, Bull. Math. Soc. Sci. Math. Roum. **14**(62), 9–13 (1970)
20. H. Brezis, F.E. Browder, Nonlinear ergodic theorems. Bull. Amer. Math. Soc. **82**, 959–961 (1976)

21. Ö. Çakar, A.D. Gadjiev, On uniform approximation by Bleimann, Butzer and Hahn on all positive semiaxis. Trans. Acad. Sci. Azerb. Ser. Phys. Tech. Math. Sci. **19**, 21–26 (1999)
22. Y.S. Chow, Delyed sums and Borel summability of independent, identically distributed random variables. Bull. Inst. Math. Acad. Sinica **1**, 201–220 (1973)
23. Y.S. Chow, T.L. Lai, Limiting behaviour of weighted sums variables, sums of independent variables. Ann. Prob. **1**, 810–824 (1973)
24. J. Christopher, The asymptotic density of some $k$-dimensional sets. Amer. Math. Monthly **63**, 399–401 (1956)
25. J.S. Connor, The statistical and strong $p$-Cesàro convergence of sequences. Anal. (Munich) **8**, 47–63 (1988)
26. R.G. Cooke, *Infinite Matrices and Sequence Spaces* (MacMillan, London, 1950)
27. R.G. Cooke, P. Dienes, The effective range of generalized limit processes. Proc. London Math. Soc. **12**, 299–304 (1937)
28. V.F. Cowling, Summability and analytic continuation. Proc. Amer. Math. Soc. **1**, 536–542 (1950)
29. J.P. Duran, Infinite matrices and almost convergence. Math. Z. **128**, 75–83 (1972)
30. O.H.H. Edely, M. Mursaleen, On statistical $A$-summability. Math. Comput. Model. **49**, 672–680 (2009)
31. C. Eizen, G. Laush, Infinite matrices and almost convergence. Math. Japon. **14**, 137–143 (1969)
32. E. Erkuş, O. Duman, $A$-Statistical extension of the Korovkin type approximation theorem. Proc. Indian Acad. Sci. (Math. Sci.) **115**(4), 499–508 (2005)
33. H. Fast, Sur la convergence statistique. Colloq. Math. **2**, 241–244 (1951)
34. N.J. Fine, On the Walsh functions. Trans. Amer. Math. Soc. **65**, 372–414 (1949)
35. R.L. Franks, R.P. Marzec, A theorem on mean value iterations. Proc. Amer. Math. Soc. **30**, 324–326 (1971)
36. A.R. Freedman, J.J. Sember, Densities and summability. Pacific J. Math. **95**, 293–305 (1981)
37. J.A. Fridy, On statistical convergence. Anal. (Munich) **5**, 301–313 (1985)
38. J.A. Fridy and C. Orhan, Lacunary statistical convergence, Pacific J. Math. **160**, 43–51 (1993)
39. A.D. Gadjiev, C. Orhan, Some approximation theorems via statistical convergence. Rocky Mt. J. Math. **32**, 129–138 (2002)
40. D.K. Ganguly, D. Bandyopadhyay, Some results on fixed point theorem using infinite matrix of regular type. Soochow J. Math. **17**, 269–285 (1991)
41. G.H. Hardy, *Divergent Series* (Clarendon Press, Oxford, 1949)
42. G.H. Hardy, J.E. Littlewood, On a Tauberian theorem for Lambert's series and some fundamental theorems in the analytical theory of numbers. Proc. London Math. Soc. **2**, 21–29 (1921)
43. G.H. Hardy, E.M. Wright, *An Introduction to the Theory of Numbers*, 4th edn. (Oxford University Press, London, 1959)
44. F.W. Hartmann, Summability tests for singular points. Canad. Math. Bull. **15**(4), 525–528 (1972)
45. B.P. Hillam, Fixed point iterations and infinite matrices, and subsequential limit points of fixed point sets. Ph.D. Dissertation, University of California, Riverside, June 1973
46. A. Jakimovski, A generalization of the Lototski method of summability. Michigan Math. J. **6**, 277–290 (1959)
47. A. Jakimovski, Analytic continuation and summability of power series. Michigan Math. J. **11**, 353–356 (1964)
48. B. Jamison, S. Orey, W. Pruitt, Convergence of weighted averages of independent variables. Z. Wahrscheinlichkeitstheorie **4**, 40–44 (1965)
49. J.P. King, Tests for singular points. Amer. Math. Monthly **72**, 870–873 (1965)
50. J.P. King, Almost summable sequences. Proc. Amer. Math. Soc. **17**, 1219–1225 (1966)
51. J.P. King, Almost summable Taylor series. J. Anal. Math. **22**, 363–369 (1969)
52. K. Knopp, Uber lambertsche reihen. J. Reine Angew. Math. **142**, 283–315 (1913)
53. K. Knopp, *The Theory of Functions,* Part 1 (Dover, New York, 1945)

54. E. Kolk, Matrix summability of statistically convergent sequences. Anal. (Munich) **13**, 77–83 (1993)
55. P.P. Korovkin, *Linear Operators and Approximation Theory* (Hindustan Publishing Corporation, Delhi, 1960)
56. T.L. Lai, Summability methods for independent, identically distributed random variables. Proc. Amer. Math. Soc. **45**, 253–261 (1974)
57. M. Loève, *Probability Theory* (Van Nostrand, New York, 1955)
58. G.G. Lorentz, A contribution to the theory of divergent sequences. Acta Math. **80**, 167–190 (1948)
59. I.J. Maddox, *Elements of Functional Analysis,* 2nd edn. (The University Press, Cambridge, 1988)
60. S.A. Mohiuddine, An application of almost convergence in approximation theorems. Appl. Math. Lett. **2**, 1856–1860 (2011)
61. S.A. Mohiuddine, A. Alotaibi, M. Mursaleen, Statistical summability $(C, 1)$ and a Korovkin type approximation theorem. J. Inequal. Appl. **2012**, 172 (2012). doi:10.1186/1029-242X-2012-172
62. S.A. Mohiuddine, A. Alotaibi, M. Mursaleen, Statistical convergence of double sequences in locally solid Riesz spaces. Abstr. Appl. Anal. **2012**, Article ID 719729, 9 pp (2012). doi:10.1155/2012/719729
63. F. Móricz, Tauberian conditions under which statistical convergence follows from statistical summability $(C, 1)$. J. Math. Anal. Appl. **275**, 277–287 (2002)
64. F. Móricz, Statistical convergence of multiple sequences. Arch. Math. (Basel) **81**, 82–89 (2003)
65. F. Móricz, Theorems relating to statistical harmonic summability and ordinary convergence of slowly decreasing or oscillating sequences. Anal. (Munich) **24**, 127–145 (2004)
66. F. Móricz, Ordinary convergence follows from statistical summability (C,1) in the case of slowly decreasing or oscillating sequences. Colloq. Math. **99**, 207–219 (2004)
67. F. Móricz, Statistical convergence of sequences and series of complex numbers with applications in Fourier analysis and summability. Analysis Math. **39**, 271–285 (2013)
68. F. Móricz, C. Orhan, Tauberian conditions under which statistical convergence follows from statistical summability by weighted means. Stud. Sci. Math. Hung. **41**(4), 391–403 (2004)
69. M. Mursaleen, Application of infinite matrices to Walsh functions. Demonstratio Math. **27**(2), 279–282 (1994)
70. M. Mursaleen, $\lambda$-Statistical convergence. Math. Slovaca **50**, 111–115 (2000)
71. M. Mursaleen, A. Alotaibi, Statistical summability and approximation by de la Vallée-Poussin mean. Appl. Math. Lett. **24**, 320–324 (2011) [Erratum: Appl. Math. Lett. **25**, 665 (2012)]
72. M. Mursaleen and A. Alotaibi, Statistical lacunary summability and a Korovkin type approximation theorem. Ann. Univ. Ferrara **57**(2), 373–381 (2011)
73. M. Mursaleen, A. Alotaibi, Korovkin type approximation theorem for functions of two variables through statistical $A$-summability. Adv. Difference Equ. **2012**, 65 (2012). doi:10.1186/1687-1847-2012-65
74. M. Mursaleen, O.H.H. Edely, Statistical convergence of double sequences. J. Math. Anal. Appl. **288**, 223–231 (2003)
75. M. Mursaleen, V. Karakaya, M. Ertürk, F. Gürsoy, Weighted statistical convergence and its application to Korovkin type approximation theorem. Appl. Math. Comput. **218**, 9132–9137 (2012)
76. M. Mursaleen, S.A. Mohiuddine, On lacunary statistical convergence with respect to the intuitionistic fuzzy normed space. J. Comput. Appl. Math. **233**, 142–149 (2009)
77. M. Mursaleen, S.A. Mohiuddine, Statistical convergence of double sequences in intuitionistic fuzzy normed spaces. Chaos Solitons Fract. **41**, 2414–2421 (2009)
78. Y. Okada, Über die Annäherung analyticher functionen. Math. Z. **23**, 62–71 (1925)
79. Z. Opial, Weak convergence of the sequence of the successive approximations for nonexpansive mappings in Banach spaces. Bull. Amer. Math. Soc. **73**, 591–597 (1967)

80. A. Peyerimhoff, *Lectures on Summability*, Lectures Notes in Mathematics, vol. 107, (Springer, New York, 1969)
81. W.E. Pruitt, Summability of independent, random variables. J. Math. Mech. **15**, 769–776 (1966)
82. S. Reich, Almost convergence and nonlinear ergodic theorems. J. Approx. Theory **24**, 269–272 (1978)
83. J. Reinermann, Über Toeplitzsche Iterationsverfahren und einige ihre Anwendungen in der konstruktiven Fixpunktheorie. Studia Math. **32**, 209–227 (1969)
84. B.E. Rhoades, Fixed point iterations using infinite matrices. Trans. Amer. Math. Soc. **196**, 161–176 (1974)
85. D.C. Russel, Summability of power series on continuous arcs outside the circle of convergence. Bull. Acad. Roy. Belg. **45**, 1006–1030 (1959)
86. A.H. Saifi, Applications of summability methods. M.Phil Dissertation, Aligarh Muslim University, Aligarh, 1991
87. T. Šalát, On statistically convergent sequences of real numbers. Math. Slovaca **30**, 139–150 (1980)
88. P. Schaefer, Matrix transformations of almost convergent sequences. Math. Z. **112**, 321–325 (1969)
89. P. Schaefer, Infinite matrices and invariant means. Proc. Amer. Math. Soc. **36**, 104–110 (1972)
90. I.J. Schoenberg, The integrability of certain functions and related summability methods. Amer. Math. Monthly **66**, 361–375 (1959)
91. A.H. Siddiqi, On the summability of a sequence of Walsh functions. J. Austra. Math. Soc. **10**, 157–163 (1969)
92. H.M. Srivastava, M. Mursaleen and Asif Khan, Generalized equi-statistical convergence of positive linear operators and associated approximation theorems. Math. Comput. Model. **55**, 2040–2051 (2012)
93. H. Steinhauss, Sur la convergence ordinaire et la asymptotique. Colloq. Math. **2**, 73–74 (1951)
94. T. Teghem, Sur des extensions d'une methode de prolongement analytique de Borel. *Colloque Sur la Theore Des Suites* (Bruxelles, 1957) pp. 87–95
95. P. Vermes, Series to series transformations and analytic continuation by matrix methods. Amer. J. Math. **71**, 541–562 (1949)
96. P. Vermes Convolution of summability methods. J. Anal. Math. **2**, 160–177 (1952)
97. P. Vermes, Summability of power series at unbounded sets of isolated points. Bull. Acad. Roy. Belg. **44**, 830–838 (1958)
98. T.L. Walsh, *Interpolation and Approximation by Rational Functions in the Complex Domain*, vol. 20 (American Mathematical Society Colloquium Publications, New York, 1935)
99. N. Wiener, *The Fourier Integral and Certain of Its Applications* (The University Press, Cambridge, 1933)
100. A. Zygmund, On the convergence and summability of power series on the circle of convergence. Proc. London Math. Soc. **47**, 326–350 (1941)
101. A. Zygmund, *Trigonometric Series* (The University Press, Cambridge, London, 1959)

# Index

M. Mursaleen, *Applied Summability Methods*, SpringerBriefs in Mathematics,          123
DOI 10.1007/978-3-319-04609-9, © M. Mursaleen 2014